MW01101974

UNIFIED FLUID DYNAMIC THEORY OF PHYSICS

A physical explanation of gravity, matter,
electromagnetic forces, photons,
the strong and weak forces,
quantum mechanics,
and much
more

THOMAS G. LANG, PH.D.

AuthorHouse™
1663 Liberty Drive
Bloomington, IN 47403
www.authorhouse.com
Phone: 1-800-839-8640

First published by AuthorHouse 8/29/2011

ISBN: 978-1-4634-4154-8 (sc)
ISBN: 978-1-4634-4155-5 (e)

Library of Congress Control Number: 2011913103

Printed in the United States of America

FOREWORD

According to modern physics, space is an empty vacuum. I find it hard to believe that gravitational, electromagnetic, and quantum fields can physically exist in an empty vacuum. It is strange that there is still no theory that unifies these fields into a physically understandable theory. Also, there are far too many phenomena in modern physics that cannot be physically understood, which I call *mysteries in modern physics*.

It was for these reasons that I started, a long time ago, to develop this new theory of physics that is based on fundamental laws, including those used in fluid dynamics. This theory physically explains gravity, matter, photons, electric and magnetic fields, quantum mechanics, and many other things; it also unifies physics, and agrees with experimental results.

I will show how this new approach to physics eliminated the thought barriers that prevented a unification of physics. The new theory that I am about to describe is remarkably different from Einstein's relativity, quantum mechanics, and the ether theory of the 1800's.

CONTENTS

1. INTRODUCTION

This book was primarily written for the public, especially for people who are interested in new ideas in science and engineering. Science is the study of things, living or not; and engineering is the application of science.

Physics is the part of science that relates to matter, energy and motion, and is the foundation of science. The theory that I am about to describe greatly strengthens this foundation by providing a *relatively simple new theory of physics, and by unifying physics*.

1.1 Background

I developed this new theory as a spare time activity that started in 1950, two years after graduating in mechanical engineering from Caltech. *I probably would not have developed this theory if I had become a physicist.*

At that time, no one knew what an electric charge is made of, what electrons and protons look like, and what physically causes electricity, magnetism, gravity, and quantum phenomena. Even after all of these years, modern physics still has no answers.

I started with the intuitive idea that *photons pulse*. This idea led to many other new concepts that together provided a physical understanding of all fields in physics, and to the unification of these fields without resorting to new dimensions in space or warping space and time.

The scope and readability of this book were greatly improved by suggestions from associates, friends, and relatives who reviewed drafts of this book. Their contributions are gratefully acknowledged in Section 14.

This new theory was originally published as "A Proposed Unified Field Theory" in three parts in *Galilean Electrodynamics* (GED) in 2000, 2001 and 2003. GED is a unique, international, peer-reviewed journal dedicated to new concepts in physics. This theory is reproduced here in the first three appendices with the permission and help of the GED editor in chief, Professor Cynthia Kolb Whitney.

My primary goal was to make this book interesting and useful to people in the general public who like new ideas, and want to learn more about science and physics.

I further wanted to answer questions that physicists and engineers might ask. To do this, I occasionally placed the more technical explanations in parentheses that other readers might want to skip. Additionally, I wanted to include theory for people with Ph.D. degrees in science or engineering, so I reproduced the original GED paper in Appendices I, II, and III. These appendices also contain the illustrations that I will be referencing for all readers. Appendix IV was added because it is a technical summary that becomes useful after reading the first several sections in this book. The combined Glossary-and-Section-Index was included to more simply describe these new ideas, and to help readers find information more quickly.

This new theory stemmed from an early interest in science that grew in high school, and expanded while I was at Caltech. After graduating from Caltech, my interest in science leaned toward aerodynamics while earning an MS degree from USC night school. My fluid dynamics specialty began in 1951 when I started a twenty-seven-year career at a Navy lab in Pasadena that later moved to San Diego. About midway through this career, I received a 2½-year scholarship for a Ph.D. degree in aerospace engineering at the Pennsylvania State University. This postgraduate education greatly expanded my theoretical ability in fluid dynamics.

Looking back, another very important aid in developing this theory of physics was my growing experience in four very different ways to approach and solve problems, which are: (1) *design*, (2) *research*, (3) *theory*, and (4) *invention*.

I did not discuss this new physics theory outside of my immediate family *until I was reasonably sure* that this theory was correct. I then surprised the other three members of a special *Philosophical Luncheon Group* by giving them a draft of this theory on Nov 10, 1991, which included the most basic concepts of this theory.

All four of us retired from the same Navy lab, and each had a Ph.D. degree. Two had a degree in nuclear physics, one had a physics degree in underwater acoustics, and I was the only engineer. We met monthly for over a decade, and never missed a meeting, as I recall, because we enjoyed them so much. These meetings stopped only after two members passed away.

After reading my physics theory, one of the nuclear physicists, Howard A. Wilcox, mentioned that this was the most interesting of our luncheons, and Eugene Cooper, the other nuclear physicist, agreed. This theory was mostly outside the field of Daniel Andrews, the third physicist, but he participated. The nuclear physicists seemed most impressed by the idea that *photons physically transform into electrons*, perhaps because they were reminded of the many quantum similarities between photons and electrons; alternatively, they might have been reminded of that famous statement "Let There Be Light" which implies that *photons come before matter*.

1.2 The Year 1900

To complete this introduction, I will take readers on a brief trip back to the year 1900, as I did in an early version of this theory. By imagining living at that time, one can better appreciate the great advancements made by modern physics and technology, and better understand how difficult it is to predict future technology.

In 1900, there were rudimentary automobiles called horseless carriages. Horses and horse-drawn buggies were the primary means of local transportation. Trains with steam engines were becoming common. There were no streamlined automobiles, freeways, subways, fast ferries, airliners, supersonic jets, helicopters or hovercraft.

There was electricity, but it had not been harnessed. There were no electric power plants, electric transmission lines, solar cells or nuclear power plants. There were no refrigerators, air conditioners, dishwashers, vacuum cleaners, electric stoves, microwave ovens, or clothes washers and dryers.

There were no telephones, TVs, radios, phonographs, computers, copiers, scanners, digital cameras, email, web sites, VCRs, CDs, DVDs, cellular phones or GPS systems.

In 1900, there were no anti-cancer drugs, heart transplants, DNA technology, x-ray machines or MRI machines. This list could go on, but the pattern is clear.

These scientific and engineering achievements were made since 1900. Modern physics played a very key role in these great achievements. Most of these achievements would not have been predicted at that time. In fact, if they had been suddenly introduced, many would have been considered magic, or even supernatural.

Likewise, it is virtually impossible to predict what will be achieved in the next 100 years, 1,000 years, million years, or one billion years. The point here is that *impossible things do happen*.

1.3 Suggestions for Reading this Book

If a reader gets "stuck" anywhere in the book, I suggest referring to the Glossary/Index, Table of Contents, Appendix IV (a summary), or simply skipping ahead.

This book is filled with surprises. The reader will find new ideas in almost every section, beginning with the section on space and time, and ending with the universe. All surprises *agree with experiment*, as they must in order to be a viable theory. Furthermore, each surprise *is consistent with other surprises*, as it must be in order to be a unified theory. What should not be a surprise is that this theory is based largely on my specialty of fluid dynamics.

2. MYSTERIES IN MODERN PHYSICS

I consider every concept or discovery in modern physics that has no physical explanation as being a mystery in modern physics.

Of the many great physicists, few were as eloquent in describing such mysteries as the late Caltech Professor and Nobel Laureate, Richard Feynman. The following quotes are from Feynman in his three-volume book, *The Feynman Lectures on Physics*, which was co-authored by Robert Leighton and Matthew Sands, and published in 1963.

> *"The mechanical rules of 'inertia' and 'forces' are wrong--Newton's laws are wrong--in the world of atoms. Here, things behave like nothing we know of, so that it is impossible to describe this behavior in any other than analytical ways." (Vol. I, Pg. 2-6)*

> *"There is no explanation of gravitation in terms of other forces at the present time." (Vol. I, Pg. 7-10)*

> *"It is important to realize that in physics today, we have no knowledge of what energy is." (Vol. I, Pg. 4-2)*

In regard to quantum theory, and wave-particle duality, he wrote:

> *"We choose to examine a phenomenon which is impossible, <u>absolutely</u> impossible, to explain in any classical way, and which has in it the heart of quantum mechanics". (Vol. I, Pg. 37-2)*

After almost a half a century, modern physics still has no answers. The following are several more mysteries in modern physics, expressed as questions, which I will also discuss and solve, together with many other mysteries:

- What do photons look like, and why do they cycle?
- What do electrons, protons, and neutrons look like?
- How are energy and matter physically related?
- What physically causes the nuclear strong and weak forces?
- What physically causes gravity?
- What physically causes quantum phenomena?
- Can physics be unified into a single, relatively simple theory?
- Did the universe really start with the *Big Bang*?

3. SPACE AND TIME

We now begin an adventure that describes my unified theory of physics. Please bear in mind that this theory, or parts of it, may or may not be correct. However, I wrote this book because I believe that this new theory is at least mostly correct, and want to share these ideas. So, get ready for an unusual, relatively simple (but not too simple) theory that unifies physics.

3.1 Space and Time in this New Theory

As mentioned earlier, this new theory blends fundamental laws of physics with fluid dynamics. The path to this theory was filled with obstacles that occasionally halted my progress for years; however, intuition, thought, and a belief in the basic assumptions eventually led to a unified theory, and this book.

Here, space is assumed normal in the sense that it is three-dimensional, and does not warp with time. Space is filled with a *spatial fluid* that will be shown to be *necessary* in order to physically understand matter, energy and motion, which together comprise physics. The density and pressure of the spatial fluid change to provide gravity; these changes also affect the speed of light, much like changes in density and pressure affect the speed of sound in air or water. Time remains normal, except it slightly speeds up with altitude in accordance with GPS results.

In other words, space and time act quite normal, spatial fluid acts much like a normal fluid; energy, mass and momentum are conserved; gravity will be shown to be an inherent property of all photons and matter; and the speed of light in spatial fluid will be shown to act much like the speed of sound in fluids such as air and water.

Alternatively, modern physics, which includes Einstein's Relativity, is based on a nearly opposite set of assumptions. In modern physics, space is an empty vacuum, everything is relative to everything else; the speed of light in space is constant; and space and time mathematically warp together to provide gravity. Additionally, Einstein found that, when matter approaches the speed of light, time slows down, length approaches zero, and mass approaches infinity, contrary to intuition, basic laws of physics, and my theory.

Overall, this new theory provides a new physical understanding of the universe, while modern physics is based more on mathematics. However, both theories tend to agree with experiment and observation.

For example, recall the so-called *proof of relativity theory in 1919* that related to a solar eclipse showing that light from a distant star is bent because, in Relativity, the sun warps space and time. Here comes a surprise. This experiment could equally well be said to prove my theory because the *bending of light* is here caused directly by gravity since gravity is an inherent property of all photons and matter, as shown later. (Author's note: Neither theory is really *proved* by this 1919 observation.)

A second example of experiments that tend to verify Einstein's *Special Theory of Relativity* are those showing that a mass approaches infinity when its speed approaches the speed of light. Here comes another surprise. My theory similarly predicts that the power needed to accelerate any mass approaches infinity when its speed approaches the speed of light. A double surprise is that *power* never reaches infinity in my theory, and instead peaks, permitting mass to exceed the speed of light.

A third example is a recent verification of Einstein's *space-time vortex* that is formed by earth's rotation. This verification was announced on May 3, 2011 by a Stanford-NASA-Lockheed/Martin team, and resulted from the NASA Gravity Probe-B satellite experiment. This experiment also verifies my theory that predicts a *spatial-fluid vortex* induced in the spatial fluid by earth's rotation.

A reader might well ask: "How can either theory be verified or disproved if both theories agree with experimental results?" As one might guess, the answer lies in the *details* of the theories and experiments, as discussed later.

Lastly, in regard to the philosophical meaning of time in either theory, and as I stated in Appendix I: Section 2:

> "*If the speed of light is zero, then an infinite time is needed for light to travel any given distance; no action can take place, and there is no meaning to time. Alternatively, if the speed of light is infinite, then action takes place simultaneously, so again there is no meaning to time. Therefore, a finite speed of light permits the sequencing of events that leads to the idea of time.*"

3.2 Spatial Fluid

The ether theory of the 1800's was based on *ether* that filled all space, *through which the earth moved*. This popular theory was supported by most physicists, but was abandoned following an experiment conducted by Michelson and Morley in 1887 that proved that the *earth did not move through any ether*.

I believe, for reasons that will become clear later, that the universe is instead filled with a unique compressible fluid that I call spatial fluid. This spatial fluid is so different from ether that I chose not to call it ether.

Although spatial fluid cannot be seen or felt, its effects can be observed and measured. Spatial fluid has many properties of common fluids such as air or water; it flows like a fluid, contains vortices (spinning regions of fluid, somewhat like whirlpools), is compressible, and follows standard fluid theory. However, the introduction of photons is new in fluid theory; therefore, I was careful to give photons properties that are compatible with both fluid dynamic theory and fundamental laws of physics. Photons are the basic elements of all rays that travel at the speed of light, such as visible light, radio waves and x-rays. Interestingly, Albert Einstein was the first to fully describe photons in a 1905 paper.

Unlike the ether theory of the 1800's, all photons and matter here consist of spatial fluid, and act like waves in this fluid. I will show that photons and matter are simply different forms of spatial fluid. There is no conflict between my theory and the Michelson-Morley experiment because the earth physically consists of this fluid, and therefore does not move or slip through this fluid, but instead the *earth moves as a part of this fluid*.

Next, I will physically describe how spatial fluid transforms into photons, what photons look like, and physically describe how they acquire momentum, spin, a moving mass, and a pulsing behavior. I will later show how photons physically lead to matter, electromagnetism, gravity, quantum effects, strong and weak forces, and many other things, including a unification of physics.

4. PHOTONS

4.1 Photons in Modern Physics

It is well known that *all photons are the same, except for their frequency*. Every photon has a certain behavioral pattern that is cyclically repeated, over and over again. Photon frequencies vary from around one cycle per second to a fantastically large number of cycles per second. For simplicity, photons are grouped into different frequency ranges.

Photons in a low-frequency range are called radio waves. Photons in a next higher frequency range are called microwaves; these kinds of photons heat food in microwave ovens. Visible light is next, where photons range in frequency from a low for red light to a high for violet light. Photons of visible light include all of the colors in a rainbow; note that red light is always at the top of a (single) rainbow. Photons in a still-higher range are called x-rays, followed by photons having the highest of all known frequencies that are called gamma rays. All of these rays, or waves, are names given to what is known in physics as *electromagnetic radiation*.

The fact that all photons are alike, except for frequency, does not mean that they are simple, but it helps in understanding them.

All photons act like both waves and particles. *When acting like waves*, photons behave somewhat like sound waves in air or water. All photons travel at the speed of light, independent of their frequency; similarly, sound waves in either air or water travel at their respective speeds of sound, independent of their frequency. Also, photons can interact with each other to produce patterns of light, much like sound waves interact to produce patterns of sound.

Alternatively, (unlike sound waves) *photons also behave like particles* because they travel in straight lines unless acted on by a force, much like billiard balls. Another particle-like behavior of photons is that they always spin around their travel axis, somewhat like rifle bullets. However, unlike bullets, experiments show that photons spin one way or the other in *exactly equal numbers throughout the universe*. Interestingly, photon spin is independent of photon frequency.

Photon frequencies cover a fantastic range. Visible light lies near the middle of all known types of electromagnetic radiation, and is centered at a frequency of 1,000,000,000,000,000 Hz, which is one thousand, trillion Hz, and can be written as 10^{15} Hz. Count the 15 zeros. (Hz is an abbreviation named after Heinrich Hertz, a physicist, meaning cycles per second.) In other words, each cycle of a photon of visible-light occurs in a very tiny fraction of a second. The frequencies of commonly known photons range from about 10^{3} Hz for radio waves, up to a fantastically high 10^{22} Hz for gamma rays.

Physicists experimentally discovered that *photon energy is exactly proportional to photon frequency*. In other words, gamma rays are immensely more energetic than radio waves.

However, modern physics offers no physical description of a photon, what physically changes during each cycle, or why photon energy is exactly proportional to frequency. These unknowns add three more mysteries to modern physics.

4.2 Photons in This New Theory

Photons in both modern physics and in this theory agree with experiments, as they must in order for each to be a viable theory. However, the descriptions of photons in the two theories are extremely different [Appendix I, Section 2]. (Author's note: Brackets are used to reference sections in the appendices.)

Here, photons are considered to be the most important things in the universe because they are the building blocks for everything else. Photons are created whenever energy is suddenly transferred into the spatial fluid by what I call disturbances. These disturbances result from chemical reactions, nuclear reactions, particle collisions, particle accelerations, and other sources. Such disturbances include fire, lightening, heat, and striking a match.

In this new theory, photons: (1) consist of spatial fluid, (2) are spherical in shape, (3) travel through spatial fluid at a local speed of light, (4) cyclically expand and contract, and (5) spin either clockwise or counterclockwise around their travel axes in equal numbers throughout our universe. Each of these features is new, except (5) which is the well-known *spinning feature of photons*. The other four features are described next.

Here, photons are created from spatial fluid, and are spherical in shape; alternatively, modern physics provides no physical description of photons, or how they are physically created. Here, photon speed varies with the density and pressure of spatial fluid, unlike in modern physics where space is empty, and photon speed cannot change. Most importantly, the cyclic heart-beat-like expansion and contraction of photons is completely missing from modern physics; I call this action the *pulsing behavior* of photons. It is this pulsing behavior that provides photons with a frequency. I will soon show why and how photon pulsing provides quantum phenomena, thereby solving a major mystery in modern physics. As stated earlier, I began this new theory with the intuitive idea that *photons pulse*.

Still another characteristic of photons in this new theory is that their passage leaves *almost* no lingering effect in the spatial fluid. This characteristic is much like the passage of a sound wave in either air or water. This property is important, as shown later, because it means that photon characteristics move with photons.

(Author's note: Due to viscosity, which is a type of friction, sound waves in air or water continuously lose a tiny fraction of their energy as they travel; however, this energy loss is so small that it leaves almost no lingering effect in the fluid.)

Lastly, it is well known that photons have momentum, which is defined as mass times velocity. Consequently, since photons move at a local speed of light, *every photon must have a moving mass*.

4.3 Photon Creation

As mentioned above, photons are created whenever disturbances cause energy to be transferred into spatial fluid. Photon frequency will logically increase with the intensity of a disturbance, while the number of photons will increase with the magnitude of a disturbance.

Since pure energy is transferred into the spatial fluid when a photon is created, then no mass is transferred. Consequently, *photon mass can come only from the surrounding spatial fluid*.

But, how are photons physically created? Imagine that a sudden disturbance causes pure energy to be transferred into the spatial fluid at a given point. This sudden input of pure energy physically causes the spatial fluid surrounding this point to expand outward in all directions, *leaving a spherical vacuum*. Since a vacuum has no pressure, the pressure of the surrounding spatial fluid will soon cause the outward moving fluid to stop, and then rush back inward to fill this vacuum. Once the vacuum is filled, the inrushing fluid has nowhere to go, except to compress the innermost fluid. Think of the incredibly high pressure that is needed to stop this inrush of fluid. Soon, the innermost fluid will compress to a maximum, and all motion will halt.

Now the fun begins, because this is when photons are created. Note that the conservation laws of physics apply throughout this new theory; these laws include the conservation of mass, energy, linear and angular momentum, and spin direction. Spin direction must be conserved because angular momentum must be conserved, as evidenced by the fact that photon spin direction is equally divided each way throughout the universe.

The conservation laws of momentum and spin together require that, in isolated regions of space, *photons must be created in multiples of four*. The reason is that if only two opposite-moving photons are created, then they must be given a mutual twist at startup to conserve angular momentum. However, this *mutual twist* makes each photon spin in the same direction relative to its forward movement. Hint: visualize the startup twist, the opposite photon travel paths, and the photon spin direction relative to each photon path. Therefore, a second pair of similar photons must be created which are given an *opposite mutual twist* in order to conserve spin direction throughout the universe, as shown in Figs. 3a and 3b. (Author's Note: All figures are in numerical order in Appendices I and II.)

Typically, large numbers of photons are created, so *starbursts* of photons usually result.

We now return to the *moment of photon creation*, consisting of a tiny sphere of spatial fluid under incredibly high pressure. Imagine this sphere to rapidly expand while smoothly morphing into four adjacent spheres (in a circular pattern) such that each spins opposite to its neighbor, much like four bevel gears that mesh. Upon reaching the speed of light, each sphere will have just separated from its neighbor, and is now a photon. The impulse given to each photon, needed to reach the speed of light, initiates its pulsing behavior. Each of these four new photons is now in an early phase of its first pulsing cycle, is spinning, and is moving at the speed of light.

4.4 Photon Pulsing Behavior

To better understand the pulsing property of a photon, it is suggested that the reader begin by visualizing a photon at its most-compressed, inner end of a pulsing cycle. Now, imagine this photon to rapidly expand. Understand that, as the photon expands, both its pressure and density must reduce. Soon, the photon's internal pressure will reduce to the normal background pressure of the spatial fluid. Visualize this photon to continue expanding because its fluid is still moving outward. Soon, expansion will stop, and the photon will have over-expanded, meaning that its pressure has reduced to below the background spatial pressure.

Consequently, the photon will begin to contract. This contraction will continue even after the photon's pressure has returned to background spatial pressure because its fluid is still moving inward. This inward-moving fluid has nowhere to go, except to compress its innermost regions. Imagine, again, the extremely high pressure that the photon must reach before all inward movement can be halted. When motion stops, the photon has returned to its starting point.

Whether or not a photon can repeat this pulsing motion forever, without losing energy, will be discussed later.

Here comes another surprise. The inner end of a photon cycle is a unique point in its cycle because it is the only part of its cycle where a photon can physically be stopped and observed; elsewhere, a photon has spread out far too much to be stopped and observed. As readers might guess, the *pulsing behavior provides a photon with its quantum properties*, as discussed later in Section 8.

4.5 Photon Wavelength, Size, and Range

We know that travel distance is speed multiplied by time. (Consequently, *a photon wavelength* is calculated by multiplying the local speed of light by its time per cycle, which in terms of seconds is "1" divided by the number of cycles per second.) A quick calculation shows that the wavelength of an average photon of visible light is about 0.3mm (or 0.01"), which is very short.

Since a photon expands out and compresses back, a photon has no specific size; however, every photon has a specific moving mass, as mentioned earlier. The minimum size of a photon, when fully compressed, is very close to zero relative to its wavelength.

When a photon expands, it also physically moves adjacent fluid outward. If spatial fluid was incompressible, then hydrodynamic theory shows that photon expansion would be immediately felt out to infinity. However, since spatial fluid is compressible, photon expansion cannot be immediately felt out to infinity. This topic is explored again later, so read on.

4.6 Photon Energy and Frequency

As first theorized by Einstein in 1905, and later verified by experiment, the *nuclear energy*, E, of any mass, m, is $E=mc^2$, where "c" is the symbol commonly used for the speed of light. This equation holds equally well here for photon moving mass.

It is well known in physics that the forward-moving kinetic energy of anything is ½ its mass multiplied by the square of its speed. Therefore, the forward-moving kinetic energy of a photon is already, by itself, equal to half of its nuclear energy. Consequently, the missing half of its energy must consist of spinning energy and pulsing energy. Note that all three kinds of photon energy are kinetic energy, which is defined as energy caused by movement. Therefore, in this new theory, *photon energy consists solely of kinetic energy*, which is *pure energy*.

It is further important to note that experiments show that photon energy is exactly proportional to photon frequency, and the multiplier is called Plank's constant.

Why is photon energy exactly proportional to frequency? In this new theory, a high-energy photon will physically expand faster than a low-energy photon because it is more highly compressed; and, it will also contract faster because expansion and contraction are theoretically similar. To physically answer the above question, I will use an analogy between photons and coil springs. Both items cyclically store and release energy, and their cyclic frequencies are calculated using similar equations. Since coil spring experiments, theory, and experience show that coil spring frequency is exactly proportional to its *stored energy*, i.e., spring strength, it is now physically understandable, by analogy, why photon frequency is exactly proportional to *photon energy*.

Summarizing, photons in this new theory can be visualized as being *moving spherical springs*. Examples of moving spherical springs in real life are underwater explosions and cavitation bubbles. Motion picture photos of underwater explosions show that an explosion bubble expands and contracts, over and over again, acting like a spherical spring, as it slowly rises to the surface. Similarly, tiny cavitation bubbles are found to cyclically expand and contract as they follow the water flow.

4.7 Photon Pressure Fields

I described in Section 4.3 how and why the mass of every photon physically comes from the surrounding spatial fluid. I then showed in Section 4.4 how and *why this same mass* is repeatedly returned to, and then taken back from, the surrounding spatial fluid, during each and every photon cycle.

This rapid, and continuing, back-and-forth transfer of mass gives every photon an effective *averaged mass* that I call *the photon mass*. Consequently, on the average, and in compliance with the conservation laws, every photon must be surrounded by a *region of spatial fluid whose mass is reduced* by this same amount. In other words, every photon lies at the center of a region of reduced spatial pressure and density.

Ignoring *density* for now, and in accordance with [I, Section 8, and Fig. 17], the *pressure drop* in this region is assumed to vary exactly inversely with distance from the photon. The amount of this pressure drop is proportional to photon mass and energy. (In theory, the density also reduces, but not necessarily at the same rate as pressure.)

Summarizing, photons are created whenever energy is transferred into the spatial fluid; this energy input creates a spherical vacuum. Spatial fluid then rushes in to fill this vacuum, forming a region of tremendous pressure. This region immediately expands, while morphing into four photons. Each of these four photons begins to spin opposite to its neighbor as it accelerates to the speed of light. This sudden acceleration causes each photon to pulse at a frequency that is proportional to its energy. On the average during each pulse, every photon acquires a mass that is removed from the surrounding fluid. Consequently, every photon is surrounded by a region of spatial fluid whose mass is less than normal, and whose spatial pressure is assumed to reduce exactly inversely with distance from the photon.

5. MATTER

The world's nuclear reactors are clear evidence that matter converts into energy. But, can energy convert back into matter?

The answer is *yes*. Fairly recent experiments with proton colliders show that energy in the form of gamma rays released during proton collisions sometimes transform into electrons and positrons, which are the lightest known particles of matter. Modern physics provides no physical explanation for this transformation, or what electrons and positrons look like. The following discussion solves these two mysteries.

5.1 Electron Rings

In this new theory, electrons are formed from photons, and are what I call *electron rings* [I, Section 3]. An electron ring forms when two high-energy gamma photons approach each other along parallel paths, fluid-dynamically attract, and then orbit each other, as illustrated in Fig. 4. These photons will then fluidly elongate along their common circular path to form what I call a *fluid ring*, as shown in Fig. 5.

But, what causes the original photon attraction? A short answer is: "If two opposite-moving photons have the same spin direction and approach each other closely enough along parallel paths, then they fluid-dynamically attract each other." This attraction force is enhanced if the photons also have the same pulsing frequency and approximately the same phase. Conservation laws show that the resulting fluid ring must rotate at the speed of light. The core of this ring will spin and pulse, just like the original photons, as shown in Fig. 5. This spinning core of fluid resembles a smoke ring, and is called a *circular vortex* in fluid dynamics.

Most readers may want to skip the following two paragraphs since they relate to technical details on electron formation.

(Two oppositely spinning spheres are well known in fluid dynamics to attract each other. Note that any two spherical photons that spin in the same direction will have *opposite relative spins* if they approach each other along parallel paths; consequently, these oppositely spinning photons will fluid-dynamically attract each other. The attraction between two *pulsing* photons is much different. A pulsing photon can be simulated by a fluid *source* when expanding, and by a fluid *sink* when contracting. A source is a fictional point in a fluid from which fluid flows in all directions. A sink is a fictional point in a fluid into which fluid flows from all directions. Either two sources or two sinks are known to fluid-dynamically attract each other; therefore, if two photons pulse at the same frequency, and are close enough in phase, then they will attract. Details on source or sink attraction can be found in L. M. Milne-Thompson's "Theoretical Hydrodynamics", 4th edition, The Macmillan Company, Pg. 211 for 2-d sources, and Pgs. 470-472 for 3-d sources. Furthermore, note that the *attraction force* between two sources should not be confused with the more-commonly-known *repulsion force* between two opposing jets. If this result seems counter-intuitive, understand that sources and jets are very different kinds of hydrodynamic flows.)

(The ring that is formed by the orbiting photons has twice the mass of each original photon. The energy-frequency law discussed earlier requires that this new ring must pulse at twice the frequency of each original photon. Furthermore, a tiny fraction of the total ring mass must transform into what is known in physics as binding energy that serves to keep the ring from breaking apart. Therefore, overall, the two photons must undergo considerable morphing since they must merge, change their orbital radius, double their frequency, and finish by pulsing exactly in phase.)

I define rings whose cores spin counter-clockwise in the direction of fluid motion as being *electron rings*, in order to agree with electron experiments and nomenclature. Rings with a clockwise core spin are *positron rings*. These two kinds of rings are identical, except for the direction of their core spins.

Note that I use the terms *the core spins and the ring rotates*. One way to remember these terms is to recall that photon spin becomes the core spin of ring cores, and that photon speed becomes the rotational speed of fluid rings.

Because photon spin is equally divided each way throughout the universe, the number of electron rings will exactly equal the number of positron rings that are formed. Since there are no natural-occurring positrons in the universe, one might well ask: "What happened to all of these positrons?" This question will be answered shortly, so read on.

The energy of electrons is well documented, and is such that the formative photons must be gamma rays with a frequency that lies near the center of the upper half of the gamma ray frequency range. This result means that electron rings are formed from the most energetic of all types of photons, and contain almost a maximum energy, which makes them especially remarkable.

The brief energy analysis that I applied to photons in Section 3.8 applies equally well to electron rings. Consequently, the at-rest nuclear energy of an electron ring is $E=mc^2$, where "E" and "m" now refer to the electron nuclear energy and electron mass, instead of photon energy and mass. Analogous to the case of photons, half of electron ring nuclear energy consists of ring speed kinetic energy, and the other half consists of core spin energy and core pulsing energy (plus a very small binding energy needed to keep the ring intact).

The physical characteristics of electron rings are in agreement with experimental measurements of electron angular momentum and magnetic moment [I, Section 4 under "Angular momentum and magnetic moment of electrons"]. This agreement is important because it supports this new theory, and is significant because modern physics has no physical description of electrons.

Summarizing, I have just shown that matter, in the form of electron and positron rings, is simply a special form of energy. One might say that "*matter is energy moving in circles*", "*matter is circles of energy*" or "*matter is harnessed energy*". Physically, the main difference between matter and energy is that matter can be held, while energy, in the form of photons, cannot be held because photons must always move at a local speed of light. Both photons and matter consist here of pure kinetic energy, which is a major surprise. Another surprise is that this kinetic energy consists of all of the three possible types of kinetic energy: linear, rotational, and pulsing kinetic energy.

5.2 Fluid Rings

In this new theory, all fluid rings, such as electron rings and positron rings, and other rings to be discussed later, replace charged particles in modern physics. Fluid-dynamic rings exhibit the following seven kinds of behaviors [I, Section 4]:

(1). If a fluid ring is isolated in spatial fluid, and is free to move, then this ring will accelerate until it reaches a "*natural ring speed*". This natural ring speed is similar to the natural ring speed of a smoke ring. As readers may know, smoke rings always move; none are stationary. Smoke rings are called *ring vortices* in fluid dynamics. Any ring vortex has a spinning core that induces a natural speed on itself that lies in the direction of the fluid flow that is induced through its center by the spinning core.

(2). A ring vortex acts like a *tiny fluid pump* if it is held in spatial fluid. This pump-like action is caused by the spin of the ring core, as shown in Fig. 7. The spinning core induces spatial fluid to flow through the inside of the ring, acting much like a propeller. Note that if this ring is released, then it will accelerate until it moves at its natural speed, as discussed in (1) above.

(3). If a fluid ring *moves sideward in its plane* through spatial fluid, or if spatial fluid moves sideward past a fluid ring, a fluid dynamic side force is exerted on the ring that is perpendicular to the oncoming flow, and in the direction toward that side of the ring which moves in the same direction as the oncoming flow, as shown in Fig. 8. This same type of fluid dynamic side force causes a spinning golf ball to *slice*, as discussed later.

(4). *Two side-by-side rings* will repel if they rotate in the same direction, as shown in Fig. 10. Note that the sides closest to each other move in opposite directions, causing the fluid flow to slow down and increase the pressure between them, pushing them apart. Alternatively, if the two rings rotate in opposite directions, as shown in Fig. 12, then the sides closest to each other move in the same direction, causing the fluid to speed up and reduce the pressure between them, pulling the rings together. These *ring-ring forces* are analogous to electric forces in modern physics.

(5). The spinning cores of two rings, when one ring is placed either *inside another ring* or is *offset sideward in a parallel plane*, will dynamically repel if the cores spin in the same direction, and will dynamically attract if the cores spin in opposite directions. (The reasons for repulsion and attraction of spinning cores are the same as in the behavior described in (4) above.) These core-core forces are thousands of times greater than ring-ring forces, as discussed later.

(6). Any freely moving ring will dynamically *point in the direction of its movement*, as shown in Fig. 9. And, somewhat similarly, if two such rings are pinned in the spatial fluid near to each other, then they will each dynamically rotate until they point in the same direction. To describe its orientation, a ring is said to *point* in the direction of the fluid that is induced through its center by its spinning core.

(7). Lastly, *all fluid rings pulse* because they inherit their properties from pulsing photons. This pulsing behavior gives matter its quantum properties. Furthermore, every pulsing ring is *surrounded by a region of reduced spatial pressure* for the same reason that every pulsing photon is surrounded by a region of reduced spatial pressure, as described earlier in Section 4.7.

Modern physics contains no mention of this seventh behavior. *This lack of a pulsing concept is precisely why I believe that quantum phenomena cannot be physically understood in modern physics*, as stated by Feynman in Section 2. Quantum phenomena are treated in Section 8.

5.3 Proton Rings

The mass of a proton is well known in modern physics to be 1836.25 times greater than that of a positron or electron, and its size is that many times smaller. Yet a proton is found to electrically behave like a positron. How can this be?

Before answering this question, I will answer an earlier question: "What happened to all of the positrons?" A short answer is: "Each positron ring, together with *many additional pairs* of electron rings and positron rings, are hypothesized here to physically transform into a proton ring. Consequently, the number of electrons in the universe will be exactly equal to the number of protons, as experimentally observed, which solves still another mystery in modern physics."

(In this theory, a proton ring is hypothesized to consist of 919 positron rings and 918 electron rings, which totals 1837, as discussed in [II, Section 7]. I attribute the difference in mass between 1837 and 1836.25 as being the binding energy of a proton ring that is needed to keep the ring intact, which is 0.041% of proton energy.)

The following is one of the simpler physical explanations that I have found for forming a proton. Other explanations are possible, such as those presented in Appendix II, referenced just above. I believe that protons can be formed only in regions where large numbers of electrons and positrons reside. Consequently, protons can form during nuclear reactions, in regions of new star formation, in the interior of stars, or as a result of proton collisions such as in a proton collider.

In any case, I believe that the formation of a proton ring begins with a single positron ring. This positron ring will fluid-dynamically attract many electron rings. As electron rings approach the positron ring, they will tend to align with it, while repelling each other. Eventually, two electron rings will migrate to opposite sides of the positron ring. For reasons presented in Section 5.2--Item (6), the three rings will dynamically interact until they lie in the same plane, *point* in the same direction, and then contact each other.

Since the positron ring is in the center, and rotates opposite to the electron rings, the three rings will mesh, much like three meshing gears, thereby avoiding annihilation. After meshing, I envision the spinning fluid to smoothly drain from the core of each electron ring into the core of the central positron ring.

Applying conservation laws, the core of this new ring will have the same spin as one electron, but will have three times its mass, 1/3 its ring diameter, and three times its frequency. This ring will continue at all times to rotate at the speed of light. *I call this ring an e-3 ring* because it acts like an electron ring because of spin, is geometrically similar, has 3 times its mass, and is 1/3 its size.

This e-3 ring will then similarly attract and merge with two positron rings to form what I call a p-5 ring. Next, this p-5 ring will attract and combine with two electron rings to form an e-7 ring, which combines with two positron rings to form a p-9 ring, and so on. (Note that an alternative, much faster way to form larger rings is for two e-3 rings to combine with a p-5 ring to form an e-11 ring, and so on. However, at least for now, I prefer the simpler idea that is described here.)

I hypothesize that this process will continue until a *p-1837 ring is formed, which is a proton ring.* This proton ring acts very much like a positron ring, has exactly the same core spin, is geometrically similar, but is 1,836.25 times more massive, and that many times smaller.

I further hypothesize that an electron ring that is attracted to this proton ring will instead orbit it at the electron ring's natural electron speed, *thus forming a hydrogen atom.* A hydrogen atom is well known in physics to consist of a proton nucleus that is orbited at a very large distance by an electron.

Now consider what might result if a proton ring annihilates. Mass, energy, momentum and core spin are conserved. Any electrons, positrons and different kinds of unstable fluid rings that might result will quickly disintegrate, resulting in a mix of photons and neutrinos that are discussed later in Section 5.5.

Note that a somewhat different result will occur if two proton rings collide head on while moving close to the speed of light. In this case, considerable kinetic energy will have been added to each proton ring, so ring particles more massive than proton rings can result. However, any more-massive particles will be unstable, so the end result will be similar to normal proton ring annihilation, except for a wider variety of more photons, and more neutrinos.

5.4 Neutrons

A neutron is electrically neutral, has a mass that exceeds a proton mass by about three electron masses, is significantly larger than a proton, and has a complex electric field, even though it is a neutral particle. Also, neutrons are known to form a very strong bond with protons within the nucleus of an atom; however, when isolated, neutrons disintegrate in about 10 to 15 minutes. Physically, what does a neutron look like, how can it bind to a proton, and why does it behave so strangely?

I hypothesize that *a neutron consists of a proton ring that lies inside an e-3 ring*, as shown in Fig. 15. In other words, both rings lie in the same plane, point in the same direction, and rotate around a common axis in the same direction.

Note that, in this geometry, the cores of these rings spin in opposite directions. Because of their opposing spins, the ring cores will attract each other with a core-core force. This physical description of a neutron solves another major mystery in modern physics.

As stated in Section 5.2--item 5, core-core forces are thousands of times greater than ring-ring forces; the reason is threefold. First, core-core distances are very short compared with typical ring-ring distances. Second, core-core forces are caused by two-dimensional (2-d) vortices, while ring-ring forces are three-dimensional (3-d) vortices, which are much weaker. Third, the strong core-core attraction force between the two rings causes the outer e-3 ring to greatly shrink, and the inner proton ring to expand slightly, further increasing their attraction force.

(Since fluid rings consist of compressed fluid, they are flexible. Ring flexibility [II, Section 7] is inversely proportional to the square root of the ring mass. Consequently, a low-mass e-3 ring is about 25 times more flexible than a high-mass proton ring. The end result is that the cores of these rings will lie quite close to each other. Calculations show that the diameter of an e-3 ring in a neutron is only 4.7 times that of a free proton.)

Interestingly, measurements of the electric field in a neutron show regions of positive and negative charge that agree with this hypothesized neutron geometry, which solves still another major mystery in modern physics.

Fluid dynamically, an isolated neutron is stable. However, random side forces caused by nearby matter, photons, or thermal energy, can move the much-lighter, outer e-3 ring far enough sideways to make it separate from the proton ring, breaking the neutron into a stable proton ring, and an unstable e-3 ring. The e-3 ring will quickly break into a stable electron ring and four gamma photons. In modern physics, a neutron is said to break into a proton, an electron, and one or two neutrinos.

5.5 Neutrinos

Neutrinos should not be confused with neutrons. Neutrinos are much like photons because they travel at the speed of light. Unlike photons, neutrinos typically pass through the earth, and are not well understood because they can be detected only by observing debris that results from their rare collisions with particles.

Comparing neutron breakup between my theory and modern physics, I conclude that a neutrino most likely consists of *two oppositely-spinning gamma rays that travel in tandem at the speed of light*, one just ahead of the other, as shown in Fig. 3b.

Such a neutrino would indeed be almost impossible to detect because of its lack of net spin, which minimizes interaction with matter. Note that three slightly different kinds of neutrinos can form, their difference being the spin direction of the leading gamma photon, and the spacing between the two gamma rays. (The spacing may either be fixed at one or more distances, or it could oscillate.) In any case, if this description is correct, then still another major mystery in modern physics has been solved.

6. OVERVIEW, FIELDS, FORCES, SPEED LIMITS

6.1 Overview

This is a good place to pause to review goals, summarize findings, discuss fluid dynamic fields and forces, and explore the speed limit of matter.

The goal of this new theory, as stated earlier, is to develop a physical understanding of physics, and to unify physics. A physical understanding means to me to be able to describe phenomena in terms of things that we know. Such descriptions include *what something looks like and why*, and *what it does and why*.

As one might guess, intuition played a key role in developing this new theory. So far, it was intuition that caused me to: (1) begin with the idea that photons pulse, (2) conceive of this new type of spatial fluid, (3) base this theory on fundamental principles, including fluid dynamics, (4) think of how photons form, (5) determine how electrons are made from photons, (6) discover how protons might form, (7) think of what neutrons look like, (8) discover that neutral particles are made from charged particle rings, and (9) determine what neutrinos are. Each of these ideas is radically different from modern physics, and yet each agrees with experimental results and solves a mystery in modern physics.

In the next sections of this book, I will describe how to physically understand nuclear physics, the strong and weak forces, quantum phenomena, gravity, electron pairing, and other phenomena. Also, I will introduce new ideas concerning the Red Shift and the Big Bang. Lastly, I will discuss the nature of our universe. The section titled "Summary, Verification and Conclusions" contains still more new ideas.

Because of the key role played by intuition in developing this theory, I decided to acknowledge intuition at the end of Section 14, which is Acknowledgments. While doing so, and based on decades of thought, I further decided to include my ideas on what intuition is, how it works, and how it can be improved, taught and learned.

6.2 Fluid Dynamic Fields

I will continue to use terms from modern physics, as needed, with the understanding that, in this new theory, electromagnetic fields, gravitational fields, and quantum fields are really fluid dynamic fields that are parts of a single overall fluid dynamic field. The two main advantages of fluid dynamic fields are:

(1) They provide a physical understanding of physics,

(2) They superimpose into a single unified field.

All fluid dynamic fields require a fluid medium, and follow conservation laws. As mentioned earlier, *photons* are new in fluid dynamic theory, so I had to introduce them carefully in order to comply with both the conservation laws and fluid dynamic theory.

By way of review, photons are formed from pure energy, have momentum, and have a moving mass that physically comes from the spatial fluid through which they travel. Also, photons have a nuclear energy, $E=mc^2$, that consists of pure kinetic energy, half of which is due to forward motion, and half is due to ring core spin and ring core pulsing.

Here, photons begin each cycle as tiny spinning spheres of spatial fluid under immense pressure. Photons then pulse outward and back, while traveling forward at the local speed of light and while spinning one way or the other about their travel axes in equal numbers throughout the universe. Their pulsing action physically provides photons with their cyclic behavior, and further provides them with their quantum behavior.

Since the pulsing action is missing from modern physics, I fully agree with Feynman in Section 2 that it is *absolutely impossible* for modern physics to physically explain quantum phenomena. The idea that *photons pulse* permits this new theory to physically explain why photons cannot be detected more accurately than one wavelength, why photons act like both particles and waves, and why photons act as though they simultaneously pass through multiple slits in a plate, as discussed in Section 8.

Furthermore, the transformation of pulsing photons into pulsing electron rings, as described in Section 5.1, shows why and how electron rings inherit their quantum behaviors from photons, and physically shows why matter contains so much energy. Also, as discussed in Section 5.3, it is now physically understandable why protons are so much more massive than positrons, and yet act electrically like positrons. Lastly, the new idea that neutrons consist of a proton ring that lies inside of an e-3 ring, as described in Section 5.4, provides a physical understanding of neutrons, describes how a neutron can bind so strongly with a proton, and shows why an isolated neutron breaks apart in 10 to 15 minutes into a proton ring, an electron ring, and neutrinos.

Readers will find many more surprises, all of which integrate into a relatively simple unified theory of physics, so read on.

Next, I will discuss how vortices are formed, and how they generate fluid-dynamic fields and cause forces.

6.3 2-d Vortices, Vortex Cores, and No-Slip Law

Vortices (spinning regions of fluid) play a key role in this theory of physics because they create fluid fields, interact with each other, and generate forces. Examples of vortices in the world around us include whirlpools, dust devils, tornadoes, and hurricanes.

In 2-d vortex theory, vortex speed falls off inversely with distance from the center of the vortex. The resulting velocity at the center of any 2-d vortex is *theoretically infinite*. However, in the real world, and in this new theory, a core exists at the center of every 2-d vortex. This core acts much like the eye of a hurricane. All such cores rotate like a rigid mass of fluid. Consequently, the speed at the center of any vortex core in this theory *is zero, and not infinity*. It follows that the maximum speed in a vortex is the fluid speed at the edge of its core.

For example, if a long spinning cylinder is immersed in water, it will soon induce a 2-d vortex in the water. The spinning cylinder is the core of this vortex. The maximum vortex speed is the peripheral speed of this cylinder.

Why doesn't the above-mentioned cylinder just simply spin, and not affect the water? A quick answer is that the water sticks to the cylinder because there is *no slip between any known solid and fluid*. Also, there is no slip within any known fluid. Furthermore, there is no slip between fluids, such as water and air, as evidenced by the fact that wind produces ocean waves. The reason for the no-slip law is that there is physically no slip between molecules of any fluid or solid. This no-slip law also applies to spatial fluid because it is assumed to flow like a fluid and follow fluid theory; therefore, adjacent pieces of spatial fluid must be connected, which means that spatial fluid follows the no-slip law.

6.4 3-d Vortices and Forces

All fluid rings in this theory have a circular core that rotates at the speed of light, spins, and pulses. Fluid rings induce two very different kinds of three-dimensional (3-d) vortices.

The first kind of vortex is a *typical 3-d vortex*, and is induced by ring rotation. The ring acts as the core of this 3-d vortex, so the maximum speed in this vortex is the ring speed, which is the local speed of light. This type of vortex is similar to the 3-d vortex induced in the air by a spinning golf ball. Note that 3-d vortices are much weaker than equivalent 2-d vortices because their sideward width is very small, and their vortex effects spread out in three dimensions.

The second kind of 3-d vortex is induced by the spinning core of a fluid ring, and resembles a smoke ring. This type of vortex is called a circular vortex in fluid dynamics. An unusual feature of a circular vortex is that it induces a natural speed on itself that is caused by its spinning core, as discussed in Section 5.2, Item (1). For example, if a smoke ring is formed at a speed that is either lower or higher than its natural speed, the smoke ring will change speed until it reaches its natural speed. An electron ring will behave similarly, unless other forces act on it.

The natural speed of an electron ring is shown in [III, Eq. 73] as being $\alpha=1/137.036$ times the speed of light, or about 1% of the speed of light. The multiplier, α, is a fundamental constant in modern physics called the Fine Structure Constant. (This constant is non-dimensional, and depends on electric charge, Plank's constant, and the so-called *permittivity of free space*.)

6.5 Electric Fields

Notice the strong similarity between electric fields in modern physics, and fluid dynamic fields in this new theory. (Also, note the similarity in their equations, as discussed in [III, Section 16].)

In modern physics, unlike electric charges attract, and like charges repel. Here, fluid rings that rotate in opposite directions attract, and rings that rotate in the same direction repel. The direction of ring rotation tends to be analogous to electric charge.

However, there are four major differences between electric fields in modern physics and equivalent fluid dynamic fields in my theory.

An important first difference is that electric fields are said to form in an empty vacuum, while fluid dynamic fields require a fluid medium.

A second fundamental difference is that no one knows what an electric charge physically consists of, or what the physical difference is between positive and negative electric charges. In my theory, an electric charge is not a thing, but instead is an *attribute* consisting of spin direction of the ring core.

A third difference is that modern physics treats electrons and protons as being point charges that have non-point attributes such as *spin up* and *spin down*, *angular momentum*, and a *magnetic moment*. Here, electron rings and proton rings are 3-dimensional objects that are physically understandable in terms of fundamental physical principles.

A fourth, and most important, difference is that electron rings and proton rings here have a pulsing property. This pulsing property is completely missing in modern physics. This pulsing property physically provides quantum and gravitational properties to both photons and matter, as discussed in Sections 8 and 9.

6.6 Magnetic Fields

The magnetic field of a bar magnet is considered here to be the fluid-dynamic flow field that results from aligning an equal number of electron rings and proton rings in a magnet, as shown in Fig. 16. These aligned rings act like tiny pumps, as described in Section 5.2--Item 2, that collectively draw in spatial fluid from the south end of the bar magnet and push it out the north end. These rings are defined to *point in the direction of the north end* of a magnet in accordance with experiments and terminology in modern physics. The return fluid flow lies outside of the magnet, and can be observed by sprinkling iron filings on a sheet of paper held above the magnet.

Within magnets, the physical effects of ring rotation cancel because equal numbers of rings rotate in opposite directions.

If either north ends or south ends of magnets face each other, the magnets will repel because opposing jets fluid-dynamically repel, independent of jet direction. Also, north and south ends of magnets attract because outward-moving jets fluid-dynamically attract inward-moving jets.

(Note that magnets can be used to provide useful force. For example, if a wire is placed perpendicular to an oncoming flow of spatial fluid generated by magnets, and if electron rings move along this wire, then a side force is produced on these electron rings, and the wire. Now, if this wire is rigidly attached to a shaft in such a way that the force on that wire causes the shaft to rotate, we then have the basic concept of an electric motor. By attaching many more wires to the shaft, we have an electric motor.)

6.8 Can Matter Exceed the Speed of Light?

A short answer is "*yes*". Note that I have already shown that spatial fluid can exceed the speed of light. For example, in Section 5.2--Item 6, I stated that an electron ring points in the direction of its motion; if this electron moves forward at speed "v", which it often does, then the fluid in its rotating ring mathematically exceeds the speed of light by a factor of the square root of $[1+v^2/c^2]$.

Note that the *speed of matter* is different from the *speed of spatial fluid*; in the above example, the speed of matter is defined as "v". Consequently, in order to show that matter can exceed the speed of light, it is necessary to show that some form of matter can exceed the speed of light; such a form could be an electron ring, proton ring, missile or even a spaceship.

Recall the similarities between the speed of sound in air and the speed of light in spatial fluid that I mentioned in Section 4.1. Since aircraft, missiles, and bullets can exceed the speed of sound in air, the similarities between these fluids suggest that matter can exceed the speed of light.

Most readers may want to skip the next four paragraphs in parentheses that technically explore this possibility.

(In standard compressible fluid theory, a critical speed exists that is defined as the speed of disturbances in a fluid. Compressible fluid theory divides into two distinct sets of problems and solutions; one set applies to speeds less than the critical speed, and the other set applies to speeds greater than the critical speed. Both sets join at the critical speed where solutions approach infinity.)

(In regard to subcritical speed solutions, engineers found a surprisingly simple multiplier that converts incompressible fluid solutions into compressible fluid solutions. This subsonic multiplier is called the *Prandtl-Glauert Rule*, which is "1" divided by the square root of $[1-(v/c)^2]$, where "v" is the speed of matter, and "c" is the critical speed, which is the speed of sound. Note that, as v/c increases from zero to one, this multiplier changes from 1 to infinity.)

(An equally simple multiplier was found for supercritical speeds that converts incompressible fluid solutions into supercritical solutions; this multiplier is "1" divided by the square root of $[(v/c)^2-1]$. Notice that, as v/c changes from one to infinity, this supercritical multiplier reduces from infinity to zero.)

(Both sets of solutions approach infinity when v=c. In the case where the compressible fluid is air, and after finding that matter can exceed the critical speed in air, engineers solved the *infinity problem* by developing a *transonic theory* that blends subsonic and supersonic theories together in a region around the speed of sound. The result showed that aircraft power increases rapidly as speed approaches the speed of sound, peaks at the speed of sound, reduces over a range of supersonic speeds, and then increases slowly as supersonic speed further increases. In other words, *nature tends to avoid infinity*.)

In his Special Theory of Relativity, Einstein mathematically found a multiplier for mass that increases from 1 to infinity as the speed of a mass increases from zero to the speed of light. Einstein concluded that it is impossible for mass to exceed the speed of light. Surprisingly, Einstein's multiplier is *identical* to the Prandtl-Glauert Rule discussed above.

Contrary to Einstein's conclusion, I believe that mass can indeed exceed the speed of light for the following reasons:

(a) Spatial fluid is hypothesized here to be a compressible fluid that follows compressible fluid theory, much like air.

(b) The *critical speed* in both fluids is defined as *the speed of disturbances*, which is the speed of sound in air, and the speed of light in spatial fluid.

(c) Both fluid media use the same *multiplier* to convert incompressible solutions into compressible solutions in the subcritical range of speeds.

(d) Therefore, *transonic air theory* should have a counterpart called *transluminal spatial fluid theory*.

(e) The fact that matter can exceed the speed of sound in air proves that its subsonic multiplier never reaches infinity. Therefore, by analogy, the equivalent subluminal multiplier in spatial fluid should not reach infinity, in which case matter can exceed the speed of light.

If matter can indeed exceed the speed of light, and if its fluid multiplier acts similar to that in air, then the power of an accelerating spaceship will peak at the speed of light, reduce to a minimum over a range of superluminal speeds, and then slowly rise as superluminal speed further increases.

Now imagine that a future spaceship is sent to a star and back, and that accurate earth-based measurements of round trip distance and time show that the spaceship made the trip at twice the speed of light. Relativity theorists would have to agree that the speed of light was exceeded, contrary to Einstein's Relativity, because both the distance and time measurements are absolute, and are not based on Relativity. However, Relativity theorists would still claim that the starship and its pilot never exceeded the speed of light because they traveled only half as far, and aged only half as much as normal. My theory would instead claim that the pilot aged normally. Therefore, if the pilot was found to age normally, then Einstein's special theory of relativity will have been disproven in each of two very different ways.

(As a totally different but related topic, one might ask: "What happens to the energy increase caused by compressibility of the spatial fluid, such as for a starship? I believe that this extra energy causes an increase in the density and pressure of the spatial fluid in the wake behind the starship, much like the result for supersonic aircraft.)

7. ATOMS

Atoms consist of a nucleus of protons and neutrons that is surrounded at a very large distance by electrons.

7.1 Structure of Atoms

The number of electrons in atoms is always equal to the number of protons, and each such number corresponds to a different atomic element. This number determines the characteristics of each of the 92 elements that are found in nature, together with the more-than-20 manufactured elements that are radioactive. For example, Element #1 is hydrogen, Element #2 is helium, and Element #82 is lead. Interestingly, lead can be transmuted into gold, which is Element #79, by removing three protons and their associated electrons; this has been done, but the cost greatly surpasses the value of gold.

The number of neutrons in the nucleus of a stable atom is known to either equal or somewhat exceed the number of protons. The sole exception is the hydrogen atom, which typically has no neutron because a neutron is not needed for reasons given shortly.

In this new theory, electron rings can dynamically orbit a nucleus at their natural speed. If electrons gain energy, then they can orbit further out from the nucleus.

Here, an orbiting electron ring always points in the direction of its movement, as mentioned in Section 5.2. Since an electron ring is a tiny gyroscope, and because its core spins counter clockwise in the direction of its movement, a gyroscopic force will cause an orbiting electron ring to continuously turn left. Therefore, an orbiting electron will never sequentially follow the same orbital path; in other words, all electron paths will lie in a spherical layer. Consequently, in view of its pulsing behavior, an orbiting electron ring will appear as a *cloudy spherical layer*, exactly as found in experiments.

(As mentioned in Section 6.4, the natural speed of an electron is "αc", which is the *base speed* of an electron in an atom, and is c/137.036, as shown in Eq. 73. Surprisingly, the radius of an electron ring is also shown to be exactly "α" times the radius of its orbit in a hydrogen atom, as in Eq. 74, where α is the well-known "fine structure constant". Geometrically, this radius ratio indicates that an atom is essentially empty. Also, since the radius of a proton ring is only 1/1836 the radius of an electron ring, the density of any atom is negligibly small compared with that of its nucleus.)

7.2 Strong Nuclear Force

Many of us were taught in high school physics that protons repel each other, neutrons have no electric charge, and that all atoms have nuclei that contain protons and neutrons. A major mystery in modern physics is: "What physically holds protons so strongly together in nuclei, knowing that protons repel each other, and neutrons do nothing?"

Experiments show that there must be a force that is thousands of times stronger than electrical forces holding protons and neutrons together; consequently, this force is called the *nuclear strong force*. Surprisingly, experiments show that this strongest force in the universe disappears outside of a nucleus. The strong nuclear force is one of the greatest *physical mysteries* in modern physics.

"What physically binds protons and neutrons so strongly together in a nucleus?" "Why does this strong force have such a short range?" "Why must neutrons exist in nuclei?"

Modern physics was unable to explain the strong force, so unobservable particles called gluons were invented that, together with other unobservable particles called quarks, bond protons and neutrons together in atomic nuclei.

Alternatively, my theory of physics provides a very simple physical explanation for the strong force. As shown in Fig. 15, a neutron consists of a proton ring that lies inside an e-3 ring; the two rings are bonded together by a core-core force. *I claim that the strong nuclear force is a core-core force.*

This new core-core force occurs only between adjacent spinning cores of fluid rings. Consequently, the strong nuclear force in this theory will physically disappear outside of a nucleus, as observed. Next, I will describe the strong nuclear force in more detail, and explain why neutrons are needed in nuclei.

7.3 Proton-Neutron Pairs

Imagine placing a proton ring on each side of an e-3 ring so that all three rings rotate in the same direction about the same axis. This is my concept of a *proton-neutron pair*, as shown in Fig. 14. This three-ring geometry is extremely stable because the core of the larger e-3 ring strongly attracts the adjacent cores of the smaller proton rings, while the cores of the smaller proton rings repel each other to keep them on opposite sides of the e-3 ring.

Each attracting and repelling core-core force is a strong nuclear force that is thousands of times greater than electrical ring-ring forces. One reason that this force is so strong is that distances between adjacent spinning cores of any two rings are orders of magnitude closer together than typical ring-ring distances. A second reason is that the two attraction forces on the more-flexible e-3 ring cause it to shrink, moving the three cores even closer together, further increasing attraction forces. And, a third reason is that the lengths of spinning 2–d cores are far greater than the thicknesses of spinning 3-d ring cores.

(As discussed in Section 5.4, an e-3 ring is about 25 times more flexible than a proton ring; therefore, the single e-3 ring will shrink about 50 times more than each proton ring expands. The end result is that the three spinning cores will lie so close together that they are bound by an immensely strong triangular set of core-core forces. The e-3 ring will end with a diameter of only 1.68 times that of a nuclear proton, as derived in Appendix II, Subsection 7, under "Elasticity of particle rings".)

It is now physically understandable what causes the strong force. It is also physically understandable why neutrons are needed in nuclei, and why the number of neutrons in stable nuclei must equal, or somewhat exceed, the number of protons. The sole exception is a typical hydrogen atom that physically does not require a neutron to stabilize its single proton in the nucleus.

Furthermore, it is now understandable why there is an upper limit to stable atomic elements; this upper limit is reached when a nucleus becomes so massive that it is geometrically impossible to strongly bind proton rings and neutrons together in a nucleus. Note that all manufactured elements have large, unstable nuclei.

7.4 Weak Nuclear Force

The weak nuclear force is thousands of times weaker than electric forces, and is experimentally found to act over a range of only one-tenth the diameter of a proton. This weak force is best known for causing beta decay, defined as the breakup of a neutron into a proton, an electron, and a neutrino.

The weak nuclear force is also known to be involved in the opposite kind of transformation where neutrons are formed from protons and either electron/positron pairs or photons.

Although the weak force seems wimpy, it literally transmutes one atomic element into another by changing the number of protons in a nucleus.

In this new theory, the weak nuclear force is made understandable by examining the structure of a neutron, which was described earlier. A neutron here consists of a proton ring that lies inside an e-3 ring. The counter-spinning cores of the two rings attract each other with a core-core force, known here as the strong nuclear force. In spite of this strong force, the geometry of a neutron is marginally stable because, with a relatively small force, the outer e-3 ring can be moved far enough *sideways* to break up a neutron. This *side force is considered the weak nuclear force*.

It is easy to understand why surrounding matter, and possibly even heat energy, can provide enough side force to break up a neutron. Intuitively, it seems that a sideward movement of around one-tenth of the diameter of a proton would be sufficient to break up a neutron, which matches the observed action distance.

7.5 Incredible Amount of Energy in Matter

It is interesting to calculate the energy stored in matter. Here, matter contains a tremendous amount of energy because it is stored at the speed of light; this already-large energy is doubled by including the kinetic energy of core spin and core pulsing. The result is that fluid rings act like *super flywheels*. I will now calculate how much energy is stored in one gallon of water.

For background, one gallon of water weighs 3,785 grams. One gram of any mass converts, via $E=mc^2$, into 24,900,000 kilowatt hours. An average U. S. household uses 10,000 kWhr/yr. Consequently, *one gallon of water* contains enough nuclear energy, if converted at 100% efficiency, to provide power to 3,785x24,900,000 / 10,000 = 9,425,000 U.S. households per year, which is almost 10 million households per year.

Note that, even at 0.1% efficiency, the need for 1,000 gallons of water is like *a drop in the bucket*. The challenge is to find a safe, low cost way to extract nuclear energy from matter. There is no other energy source in the universe that even comes close to the potential of nuclear energy to generate power at near-zero cost.

8. QUANTUM MECHANICS

Quantum mechanics is considered to be the most mysterious part of modern physics. Quantum mechanics includes the discreteness of energy and matter, uncertainty principal, action at a distance, and wave-particle duality of energy and matter. According to Feynman, as quoted in Section 2, quantum phenomena are absolutely impossible to physically understand in modern physics.

8.1 Heisenberg's Uncertainty principle

Heisenberg's uncertainty principle is fundamental in quantum mechanics. Basically, this principle means that energy and elementary particles are discrete items whose positions *cannot be measured any closer than one wavelength.*

In section 4.4, I explained this discreteness, and physically described why the position of photons cannot be measured any closer than one wavelength. Due to its pulsing property, a photon physically expands outward into the surrounding spatial fluid during each cycle, *making detection impossible* except once every wavelength when it is in its most-compressed state. All ring particles, and all matter, inherit this characteristic.

(Additionally, Heisenberg stated that: (a) the minimum uncertainty in measuring momentum is proportional to "h" divided by the minimum uncertainty in measuring position, where "h" is Plank's constant, and (b) it is impossible to measure the minimum uncertainty in both position and momentum of a photon or elementary particle at the same time. My theory further explains both (a) and (b). Plank's constant is defined as energy-divided-by-frequency which is mc*c/frequency, which is momentum times wavelength. Since the minimum uncertainty in measuring the position of a photon or particle of matter is one wavelength, then the minimum uncertainty in measuring momentum is Plank's constant divided by one wavelength. Furthermore, the uncertainty in the measurements of *position* and *momentum* relate to two very different kinds of physical measurements that are mutually exclusive. For example, the measurement of *position* results in the destruction of a photon or the halting of a particle. Alternatively, *momentum* is measured by detecting speed, direction and mass; this measurement can be accomplished only in ways that are not compatible with measuring position.)

8.2 Action at a Distance

Action-at-a-distance is a quantum phenomenon that is sometimes called a spooky phenomenon because it relates to interactions between entangled photons separated by large distances that occur far faster than the speed of light. (Entangled photons are two photons that result from splitting a parent photon.)

In my theory, action at a distance is physically possible because the pulsing property of photons and elementary particles permits their frequency, spin, and location to theoretically be detected over large distances, far faster than the speed of light [II, Section 15].

8.3 Wave-Particle Duality

It is well known that photons and elementary particles follow straight paths unless acted upon by a force; this behavior is particle-like. Alternatively, these entities sometimes act like waves, which is puzzling, and is called *wave-particle duality*.

For example, *experiments indicate that a single photon can pass simultaneously through two or more slits in a specially designed plate called a diffraction grating.* If a large number of photons are fired either singly, or all at once, at such a grating, they form a diffraction pattern on a detection plate located behind the grating. This pattern consists of regions where many photons land, and other regions where no photons land.

Fortunately, quantum physicists are able to accurately calculate the probability of a photon landing at any point on the detection plate. They do this by calculating the probability of a photon to follow each of an infinite number of possible paths that would end at every possible landing point on the detection plate.

Of course, a photon does not have the ability to calculate an infinite number of paths to an infinite number of destinations before selecting its path. Consequently, diffraction plate experiments represent another major mystery in modern physics.

My theory solves this problem by explaining how a photon or elementary particle can move, cycle by cycle, from its release point to its final destination, as it physically must do, without using probability, as described in [III, Section 10].

A reader might now ask: "How can a photon pass simultaneously through multiple slits in a diffraction grating?" A short answer is: "During each pulse, a photon expands outward in all directions, and then compresses back to a new collapse point that is one photon wavelength further along its path. During each outward and inward pulse, some spatial fluid will pass each way through each slit; therefore, physically, each slit affects the position of each subsequent photon collapse point. One can say that fluid associated with a photon passes through all slits."

Most readers may want to skip the next three paragraphs because they describe in detail how I believe a photon moves.

(Consider a photon that is launched toward a diffraction grating that contains several slits. During each of its tens of thousands of cycles, this photon will expand and contract. The maximum expanded radius of a photon, according to my intuition, is greater than one wavelength, but much less than its original distance from the grating. Note that fluid outside of the photon will also cyclically expand and contract. In other words, a photon acts like a moving spherical spring that expands and contracts as it moves forward. The fluid that lies within the photon is never the same, which is true for all waves in fluids. A photon's path can be visualized as a series of finite dots where each dot represents its most-compressed state. Each finite-sized dot will have an angular momentum caused by the photon spin. This angular momentum tends to keep a photon on a straight path, unless a side force acts on the photon; typically, side forces are caused during expansion-and-compression cycles because the slits in the grating, and the laboratory surroundings, are unsymmetrical.)

(When a photon is within a few wavelengths of the slits, the slits will more strongly affect its path. Changes in path depend on slit geometry, and especially on the ratio of slit width to photon wavelength. In one extreme case where slit width is a tiny fraction of photon wavelength, there is a high probability that the path of a photon will intersect the grating plate and stop. However, if a photon's path carries it through one of the slits then its path will be strongly affected not only by that particular slit, but by the locations and thicknesses of the other slits. In the other extreme case where slit width is much greater than a photon wavelength, the photon will have a far greater probability of passing through a slit, and its path will be relatively much less affected by other slits. In any case, I think that a *fringe effect* is possible that permits a photon to *appear to pass through the diffraction plate* in a small region adjacent to either side of any slit where two consecutive collapse points could conceivably straddle the diffraction plate.)

(Summarizing, the fluid associated with a pulsing photon whose dotted path carries it through one slit will have passed back and forth through all slits many times on its way to and from a detection plate. By definition, this photon will have interfered with itself. Note that the detailed path of such a photon is not random, but instead depends on precise values of launching angle, wave phase at launch, grating geometry, slit details, laboratory shape and size, equipment location, position of observers, etc.)

The above descriptions show why a large number of single photons fired separately in random directions toward a diffraction grating can form a diffraction pattern on a detector plate. Similarly, this theory shows that a large number of photons that are randomly fired in different directions at the same time toward the diffraction plate will form the same diffraction pattern. However, in each case, note that none of the photon paths are random; it is only their firing directions and their pulsing phase at launch that are random.

In other words, the idea that *photons pulse* permits a physical understanding of diffraction grating experiments. Consequently, this new pulsing behavior solves a major mystery in modern physics. Furthermore, it is now understandable why probability-based quantum theory works so well, but cannot provide a physical explanation.

Lastly, it is important to note that diffraction plate experiments using different slit geometries show that electrons, and even small molecules, behave much like photons, and form diffraction patterns. This result is physically explained here because matter inherits its *pulsing properties* from photons.

9. GRAVITY

9.1 Concept of Gravity

Gravity is a universal force that causes masses to attract each other. But what physically causes gravity? Modern physics has no answer.

Here, gravity is a *pressure force* [II, Section 8] that acts on any mass that lies in a *pressure gradient*. In this theory, pressure gradients are literally everywhere because every photon and particle of matter obtains its mass from the surrounding spatial fluid. As a consequence, every mass lies at the center of a region of reduced pressure, as schematically shown in Fig. 2, and described in Sections 4.7 and 5.2--Item 7. Each such region causes every other mass in the universe to be attracted to it.

In other words, every mass in the universe is inherently attracted to every other mass in the universe, which includes the mass of all photons and matter. Most readers may want to skip the following three paragraphs.

(Any pressure force acting on a mass is calculated by multiplying the volume of that mass by the local pressure gradient. But, what is meant in this theory by volume and by pressure gradient? *Volume* is defined as mass divided by density. Therefore, volume is directly proportional to mass if density is assumed to be a fixed *background fluid density* such as the average spatial density in the region over which gravity acts. *Pressure gradient* is defined as the change in pressure per unit distance. Since *pressure* at any point is assumed to reduce inversely with distance from any mass, the pressure gradient at that point *mathematically reduces inversely with the square of distance*.)

(For example, the pressure gradient at any mass M2 caused by any mass M1 is proportional to $M1/d^2$, where "d" is the distance between the masses. The gravitational force exerted on M2 is in ratio to this pressure gradient multiplied by M2, which is $M1*M2/d^2$. Exchanging M1 and M2 provides the same force, which means that each mass is pushed toward the other mass with the very same force. This result is the well-known law of gravity.)

(Similarly, the total gravitational force, F, that acts on any mass, M1, is an experimentally-obtained gravitational constant "G" multiplied by M1 times the sum of every other mass M2 in the universe divided by the square of its distance from M1, which can be abbreviated as $F=G*M1*\sum[M2/d^2]$.)

Gravity is now physically understandable as being a pressure force that causes every mass to be attracted to every other mass.

To illustrate gravity, imagine that the earth is isolated in space. Then the gravitational attraction between a person standing on the earth, and the earth, is "G" multiplied by the mass of that person, multiplied by the mass of the earth, divided by the square of that person's distance from the center of the earth.

Since pulsing, in this theory, is a quantum behavior that leads to gravity, then gravity is a quantum phenomenon. Furthermore, since experiments show that quantum phenomena occur far faster than the speed of light, then gravity most probably occurs far faster than the speed of light. But, what do experiments show?

9.2 Experiments on Gravity Speed and Range

Experimental data on the effects of gravity analyzed by the late astronomer and celestial dynamics specialist Dr. Thomas Van Flandern show that gravity acts at least 2x10^10 times faster than the speed of light (Physics Letters A 250:1-11, 1998).

Data gathered in modern physics on the range over which gravity acts typically show that gravitational forces extend outward well beyond nearby galaxies. It is believed by some physicists that the range of gravity could be infinite.

Clearly, more experimental data are needed to obtain more precise values for the speed and range of gravity. Data suggest that the speed and range of gravity might both approach infinity.

9.3 Theory on the Speed and Range of Gravity

If a fluid is incompressible, the effects of fluid motion are theoretically felt instantly out to infinity. However, if a fluid is compressible, like air, then we know that the effects of fluid motion cannot be instantly felt out to infinity.

However air, at speeds as low as 5% of the speed of sound, acts almost as if it was incompressible. Therefore, it is important to determine the compressibility of spatial fluid, and how small spatial fluid speeds must be in order for spatial fluid to act incompressible. Many readers may want to skip the next two paragraphs that technically explore these questions.

(To greatly simplify this analysis, assume that a photon expands out to a maximum radius, and contracts back from that radius, at the speed of light. This maximum expanded radius of this photon is then half its wavelength, λ. If we further assume that the fluid speed outside this photon reduces inversely as the square of distance from its center, as in incompressible flow, then the fluid speed at any non-dimensional radius, R/λ, is $v = c/(2R/\lambda)^2$. For example, if $R/\lambda=50$, then the fluid speed is $v = c/(100)^2$, which is 0.01% of the speed of light, which is very small.)

(Now consider an average photon of visible light whose wavelength is 0.3mm. At a radius of 50 wavelengths, for example, which is 15mm (0.6”), the fluid speed will be 0.01% the speed of light. Additionally, consider an electron whose wavelength is about 10 million times less than this photon, and a proton whose wavelength is two thousand times less than that of an electron. It is easily seen that, at a distance of only 15mm, the already-small fluid speed for this photon becomes millions and billions of times smaller, respectively, for electron rings and proton rings.)

As the result of the above greatly-simplified analysis, we find that fluid speeds resulting from pulsations of either photons or ring particles are so incredibly small relative to the speed of light, that the speed and range of gravity could approach infinity. However, as mentioned earlier, more data are needed on the speed and range of gravity.

10. ELECTRON PAIRS AND ORBITS

10.1 Electron Pairs

It is known that electrons pair, and form orbits. But, how can electron rings pair? Fig. 12 shows how two closely spaced electron rings can orbit each other if they lie in the same plane and rotate around a common axis that lies midway between them in that plane. This pairing resembles a rotating barbell whose end weights represent electron rings that move in opposite directions. These electron rings will attract each other with a ring-ring force because their rings rotate in opposite relative directions. This attraction force balances their centrifugal forces.

A second kind of an electron pair is theoretically possible, as shown in Fig. 11 where two electron rings lie very close to each other in offset parallel planes while pointing in opposite directions (either way) and rotating about an axis that lies midway between them. Since these rings lie in parallel planes, and point in opposite directions, their very-closely-spaced cores will spin in opposite directions. The resulting core-core attraction force between the rings counteracts their centrifugal forces. Since core-core attraction forces are thousands of times greater than ring-ring forces, their rotational rate and stored energy can be astonishingly high.

Even a third kind of electron pairing may be possible, unless it is prevented by electron pulsing or by electron ring rotation. In fluid dynamics, two *ring vortices* can theoretically pair by periodically changing fore-and-aft positions wherein the following vortex sequentially passes through the *inside* of the leading vortex, as described in many texts on fluid dynamics. This strange *leap frog action* is said to theoretically continue forever.

10.2 Electron Ring Orbits

Similar to the first type of pairing discussed above, many electron rings can theoretically orbit in a circular path around a central axis, much like horses on a merry-go-round. Each electron ring is attracted to electron rings on the opposite side of the circular path; also, each ring is slightly repelled by nearby electron rings on the same side of the path. These ring-ring forces counteract centrifugal forces, and tend to equally space the electron rings around the circular path. As discussed earlier, each electron ring *points* in the direction of its motion; and since each is a tiny gyro, each ring (and the entire orbiting group) will continuously *turn left*. The resulting paths of the pulsing electron rings will appear as a fuzzy spherical shell.

Any such set of orbiting electron rings can also orbit an atomic nucleus; however, the orbit diameter will be smaller because of the increased attraction force caused by proton rings in the nucleus.

Furthermore, two parallel sets of orbiting electron rings can orbit a nucleus; the opposing forces between the parallel sets of electron rings serve to keep the orbits apart, and off center. Also, additional sets of orbiting electrons can orbit at other radii.

10.3 Superconductivity

Pairs of electrons, called Cooper pairs, are experimentally found to be involved in superconductivity. Therefore, either of the first two kinds of electron pairing discussed above are conceivable candidates for Cooper pairs because they have no net ring rotation or net core spin, and will have almost no interaction with matter.

11. RED SHIFT AND BIG BANG

11.1 The Red Shift

It is well known that the frequency of light received from distant stars and galaxies reduces in proportion to distance. This reduction in frequency toward the red end of the light spectrum is called the Red Shift.

The Red Shift is explainable by either a Doppler shift (a velocity effect caused by stars and galaxies moving away from the earth at speeds that are proportional to distance), or by a loss in photon energy that increases with distance. However, in modern physics, the Red Shift can be caused only by a Doppler shift because space is assumed to be a vacuum, so there is no way for photons to lose energy, and no place for any lost energy to go.

11.2 Big Bang

Modern physics describes the expanding universe as being somewhat like an expanding balloon where dots placed on the balloon move away from each other at a speed that is proportional to their distance apart. Consequently, it is concluded that the universe started with a Big Bang. The idea is that all energy and matter in the universe was initially compressed into a tiny dot that suddenly expanded to become what is known as the Big Bang expanding universe. Calculations indicate that our Big Bang universe is 13.7 billion years old.

However, many astronomers are not pleased with the Big Bang theory because their studies indicate that the observed clustering of galaxies in the universe gravitationally requires far more billions of years to form than the age of the Big Bang.

In this new theory, the Red Shift can be caused by a Doppler shift and/or a photon energy loss with distance.

Consequently, here, the Big Bang either happened as currently believed, it never happened, or the universe is a modified Big Bang.

11.3 Photon Energy Loss Concept

I especially like the concept of a photon energy loss with distance for the following reasons:

(1) It can satisfy astronomers who believe that the universe should be much older than currently believed.

(2) It completes symmetry in my theory because it permits photons to convert back into spatial fluid.

(3) It satisfies an intuitive feeling that photons cannot travel forever without losing energy.

In regard to symmetry, I have shown how spatial fluid can transform into photons, how photons can transform into matter, and how matter can transform back into photons. Missing, is the transformation of photons back into spatial fluid. I will further discuss photon energy loss at the end of the next section.

12. THE UNIVERSE

12.1 Our Universe in Modern Physics

Based on the Big Bang theory, and observation, astronomers believe that there are as many as 300 billion stars in our galaxy, and around 300 billion galaxies in the universe. By multiplying these numbers, there may be about ten billion, trillion (10,000,000,000,000,000,000,000 $= 10^{\wedge 22}$) stars in the universe. However, even this number may be low if our universe did not result from the Big Bang. I will now present background information on what is known about the stars in our universe.

12.2 White Dwarfs, Neutron Stars, Black Holes

Astronomers find that the universe is a living thing in the sense that stars are continuously being born, live a long time, and then die quickly in one of three ways while leaving remnants that can last for a very long time. During their lives, stars act somewhat like matter factories because iron, carbon and a few other elements are known to form in their interiors.

Star death is a relatively fast process that begins when a star runs out of hydrogen fuel, and can no longer fuse hydrogen into helium. Dying stars typically collapse and then explode, spreading much of their matter into the universe.

The remnants of dying stars fall into three categories, called white dwarfs, neutron stars and black holes. White dwarfs have masses that are less than 1.4 times that of our sun. Neutron stars have masses lying between 1.4 and 3 times that of our sun. Black holes have masses that exceed 3 times that of our sun, and can be incredibly large, such as those called *supermassive black holes* that reside in the centers of galaxies, and have masses that are millions of times greater than our sun.

When dying, stars that become white dwarfs become unstable, throw off their outer layers, and collapse into condensed matter that consists of atomic nuclei and non-orbiting electrons. White dwarfs have densities of hundred of tons per cubic inch. White dwarfs shine dimly until all excess electron energy is shed, which takes longer than the age of a Big Bang universe.

Stars that become neutron stars die far more dramatically. Their greater gravity causes them to collapse, and then violently explode as supernovas when they run out of hydrogen fuel. These stars lose over half their mass during a supernova, while shining brighter than an entire galaxy of stars for several weeks or months. They then collapse into extremely compact masses consisting mostly of neutrons that have a density of around 4 billion tons per cubic inch, which approaches the density of atomic nuclei. A type of neutron star called pulsars rotate at speeds between 0.2 and 600 rotations/sec, which is incredibly fast for such large masses.

Stars that become black holes may or may not explode as they collapse into a mass that is so compact that their gravity prevents photons from escaping. Interestingly, black holes retain their gravitational properties; this fact supports an earlier finding here that the speed of gravity is far greater than the speed of light, and this may be why black holes exhibit gravity.

My theory has little to add or to change regarding white dwarfs or black holes. However, neutron stars deserve additional attention. Neutron stars are said to consist of a very dense core of neutrons with a sprinkling of protons and electrons, surrounded by a thin layer of atomic nuclei through which electrons flow. I think that both the reported *sprinkling of protons and electrons* in the core, and the *thin layer* would more likely consist of proton-neutron pairs and electron rings. These differences from modern physics are minor, but are worth mentioning.

12.4 Birth of New Stars and Galaxies

Scientists know that new stars and galaxies are continuously being formed throughout the universe, and believe that they result from aggregations of space dust that consists of small particles of matter and hydrogen. These aggregations appear throughout the universe, and are thought to result from stellar explosions and other sources, including star remnants. In my theory, these aggregations result mostly from a much different source, which is discussed next.

12.5 Creation of New Matter

If the Red Shift is caused at least partly by photon energy loss, then the universe will eventually disappear. While this result is not necessarily good or bad, I want to introduce an idea that prevents the universe from disappearing.

I hypothesize that, whenever a *sufficiently large region* of spatial fluid reaches a certain *maximum density*, then the spatial fluid in this region steadily transforms into gamma photons until a *minimum density* is reached. This new concept is analogous to a thermostat that keeps the room temperature between preset maximum and minimum temperatures.

As discussed in Section 5, gamma photons can transform into electron and positron rings, which in turn can lead to atoms such as hydrogen. Any such new matter would tend to collect into aggregations such as those mentioned in the previous section, and that are believed to form new stars and galaxies. If this hypothesis is correct, then the *sufficiently large region* is likely to be galactic in size. Candidates for such regions are those that tend to be the least populated with galaxies, or those that contain the oldest galaxies.

12.6 Re-Examining the Universe

The idea that photons lose energy with distance has far-reaching ramifications, including the probability that our universe is not the Big Bang as currently believed. I will show that our universe could be one of an infinite variety of either Big-Bang-type universes, or finite steady-state universes. Our universe could also be an infinite universe.

Philosophically, the beginning of any kind of universe will always be speculative. The reason is that *something must exist prior to the formation of any universe*.

Since something must exist ahead of any universe, I like the idea that space has forever been filled with spatial fluid. I favor this idea because it permits our universe to begin with something (a spatial fluid) that is almost nothing (a space without structure), and that provides everything needed to form a universe at any time.

The simplest universe that might result, but not necessarily the best, is an infinite universe that is fundamentally stable. In this universe, the Red Shift can be caused only by a photon energy loss with distance. It follows that new stars and galaxies must continually form in order to replace old stars and galaxies. Such a universe could begin at any specific time by introducing the *thermostat idea* discussed in Section 12.5. Alternatively, this universe could have existed forever, in which case the startup problem is minimized, but not eliminated.

A second kind of possible universe is finite, steady state, and very probably spherical in shape. Any number of such universes can exist in an infinite space; there is no need to invent new spatial dimensions in order to introduce parallel universes. However, each such finite universe requires a stable outer boundary. One such boundary would naturally form if a universe had sufficient mass and density to gravitationally prevent photons from escaping. A much different outer boundary is formed if photons simply disappeared by losing energy with distance; in this case, all lost energy would transform into new stars and galaxies. Any such universe could begin at any specific time by once again applying the *thermostat idea* discussed in Section 12.5, but limiting it to a specific region in space.

Alternatively, an infinite number of types of expanding universes are possible, depending on the fraction of the Red Shift that is caused by photon energy loss. This fraction ranges from zero to one. The current Big Bang concept represents the end case where the Red Shift is purely a Doppler Shift. Of the other possibilities, I intuitively prefer that type of an expanding universe where the photon energy loss, expansion rate of the universe, and the rate of new star and galaxy formation, all balance to provide what could be called a *steady-state expanding universe*. Note that any of these expanding universe can begin at any time by converting a region of spatial fluid into pure energy, and then suddenly re-introducing this energy into the center of that region. The result would be very much like the beginning of the hypothesized Big Bang.

Of the three basic kinds of universes, I currently favor a steady-state spherical universe for our universe. However, much thought is needed to solve this new puzzle of the universe, which I will leave as an exercise for the readers, at least for now.

Comments on Dark matter and dark energy

Observable, luminous matter is said to comprise only about 10% of the matter in the universe that is needed to gravitationally account for the motions of stars within galaxies, and the motions of galaxies within galactic clusters. In other words, 90% of the gravitational force that is believed needed to explain these movements consists of non-luminous matter, called dark matter.

Dark matter includes black holes, asteroids, comets, matter ejected from supernovas and other stellar explosions, aggregations of space dust and particles, hydrogen and other gasses, and cosmic rays. In this new theory, photons and neutrinos will act like dark matter. There may be so many kinds of things in the universe that do not shine, that there may be no *missing* dark matter.

Alternatively, the hypothesized loss of photon energy with travel distance, together with the resulting new regions of high-density spatial fluid, if they exist, should act like dark matter. Furthermore, any of the hypothesized new types of universes mentioned above are so radically different from the current Big Bang universe concept that the dark matter problem disappears.

As a somewhat related topic, recent measurements of the Red Shift indicate that the universe is expanding at an increasing rate. As a consequence, the idea of *dark energy* was invented, that acts like a negative gravity, to explain this finding. However, if the Red Shift is at least partially caused by a photon energy loss with distance, as I believe, then the universe will be so different from the Big Bang universe that the dark energy problem goes away.

Re-examining Time

This new theory shows that density and pressure changes in the spatial fluid are caused by the presence of photons and matter. These changes cause variations in the speed of light, the speed of time, and gravitational force.

As I mentioned earlier, the idea of time seems to be related to a finite speed of light. I think that it is possible that further research will show that the rate of time is *exactly proportional* to the speed of light, and that any relationship between gravity and time is incidental. If this is so, then a new understanding of time may have been achieved.

13. SUMMARY AND CONCLUSIONS

13.1 Summary

A new theory of physics is proposed that unifies physics, provides a physical understanding of physics, and agrees with experimental results. This theory is radically different from relativity theory, quantum theory, and the ether theory of the 1800s, as it must be, in order to solve the many physical mysteries in modern physics. The power of this new theory is that it is understandable, logical, and results from fundamental principles.

According to modern physics, it is impossible to physically understand many basic aspects of physics. If this new theory of physics is at least mostly correct, then it has indeed done the impossible. However, such an achievement is not unusual in physics, and science in general. In fact, most of the scientific discoveries and technical achievements that exist today might have been considered impossible in 1900.

For example, imagine what a person on horseback in 1900 would have thought if a jet aircraft suddenly flew by closely overhead at supersonic speed, followed a split second later by a thunderous sonic boom.

One might say, with tongue only slightly in cheek, that *achieving the impossible is seldom an impossible impossibility*.

The more unusual aspects of this new theory are:

1) The universe is filled with a unique spatial fluid.
2) Fluid dynamic fields replace all fields in physics.
3) Physics is unified into a single fluid-dynamic field.
4) Fundamental theory dominates, and space is absolute.
5) Photons pulse, and are a special form of spatial fluid.
6) Photons are created in multiples of four.

7) The speed of light varies with spatial fluid properties.
8) Matter is created from photons.
9) Matter inherits its pulsing properties from photons.
10) Matter is a special form of spatial fluid.
11) Matter is energy moving in circles.
12) Photons and matter consist solely of kinetic energy.
13) Spatial fluid, photons, and matter are interchangeable.
14) All elementary particles consist of fluid rings.
15) Fluid rings are geometrically & dynamically similar.
16) Fluid rings rotate at the speed of light, and pulse.
17) Electron and positron rings have opposite core spins.
18) Proton rings result from electron and positron rings.
19) An e-3 ring acts like an electron with 3x its mass.
20) A neutron is a proton ring lying inside an e-3 ring.
21) All neutral particles are combinations of fluid rings.
22) Ring flexibility reduces as the square root of mass.
23) Ring-ring forces are analogous to electric forces.
24) The strong nuclear force is a core-core force.
25) The weak nuclear force is the neutron breakup force.
26) Neutrinos are tandem contra-rotating gamma photons.
27) Proton-neutron pairs are 2 protons and an e-3 ring.
28) Orbiting electron rings form cloudy spherical shells.
29) Photons and matter cause spatial pressure gradients.
30) Gravity results from gradients in spatial pressure.
31) All quantum phenomena result from pulsing.
32) Quantum phenomena occur far faster than light.
33) Gravity is a quantum phenomenon.
34) The speed and range of gravity approach infinity.
35) Two kinds of electron ring pairing are possible.
36) Electron rings can orbit nuclei, or independently.
37) The Red Shift is partly caused by photon energy loss.
38) Matter can exceed the speed of light.
39) The universe is different than currently believed.
40) The Big Bang is either older, or it never happened.

13.2 Verification

A theory can be verified in at least three ways. A first way is to make a prediction based on the theory, and then verify it by conducting an experiment. A second way is to show that newly discovered phenomena are explainable by the theory. And a third way is to show that the theory physically explains existing experimental findings better than other theories.

The more unusual aspects of this new theory can be considered as predictions, and then experimentally tested, as in the first type of verification discussed above. Forty of these *more unusual aspects* are listed above. Of these, I think that the most testable are: photon pulsation, photon formation in groups of four, electron and proton ring structure and pulsation, the existence of e-3 rings, the proposed structure of neutrons, the hypothesized explanation of gravity, the idea that matter consists of circles of energy, the hypothesized proton formation process, core-core forces, ring-ring forces, photon energy loss with distance, and star and galaxy formation.

The acceptance of this new theory will surely depend upon how well it unifies physics, agrees with experiment, physically explains known phenomena, predicts or agrees with newly discovered phenomena, and solves the many physical mysteries that remain in modern physics.

13.3 Conclusions

By using fundamental principles, introducing photon pulsing, and adding many new ideas, this new theory has physically solved some of the most fundamental mysteries in modern physics. I firmly believe that a continuing pursuit of a physical understanding of physics will solve the remaining mysteries in physics, and will further help to solve mysteries in other sciences, especially chemistry and biology.

Looking back, it is surprising that it is possible to develop a radically new, futuristic unified theory of physics, such as this one, by using old and established laws of physics. One might say that this new theory goes "*Back to the Future*".

Looking ahead, history shows that any new theory or discovery that provides a better physical understanding inevitably leads to new discoveries and achievements. The discovery of DNA is a prime example.

In this theory, new physical understandings were developed for the following *groups of phenomena*:

- Electricity, magnetism, gravity, quantum mechanics.
- Photons, electrons, protons, neutrons, neutrinos.
- Electric, magnetic, strong, and weak forces.
- Spatial fluid, photon, and matter transformations.
- Fluid rings, ring combinations, ring pairs and orbits.
- Speeds of gravity, matter, and quantum phenomena.
- Red Shift and photon energy loss with distance.
- Formation of new stars and galaxies.

Here are a few general predictions of what might result from these new understandings. The first that comes to mind is a radically new type of nuclear power plant that safely generates power for a tiny fraction of current costs while leaving no radioactive waste. Unprecedented advancements in nearly every industry would follow, leading to a worldwide leap in the standard of living. Although such an advance could be based on either hot or cold, fusion or fission, I intuitively believe that the simplest approach is to combine electron rings and positron rings. The further possibility of developing small nuclear power sources for automobiles and homes would revolutionize these industries.

Other possible new developments include advanced types of: chemical compounds, metals, plastics, batteries, superconducting materials, computers, medicines, medical equipment, agriculture, communication, construction, mining, housing, and transportation.

14. ACKNOWLEDGMENTS

Many people helped to improve the content and readability of this book, for which I am most grateful.

I want first to acknowledge my parents, Glenn Irving Lang and Edna Hamann Lang who, although they had no science background, provided encouragement that started in 1950 after reading an early description of this physics theory.

I especially want to thank my wife, Patricia Stannard Lang, who I married in 1962, and who accompanied me on many weekend and weeklong retreats to work on this physics theory. With a master's degree in Education, she edited numerous drafts of both the GED paper and this book.

My greatest thanks go to my son, James Thomas Lang, who, with a master's degree in mechanical engineering from Stanford University, reviewed drafts of both the GED paper and this book, asked critical questions, offered many technical suggestions, and recommended that I add several new topics to this book.

I thank my sister-in-law Janet Stannard Aird, who has a degree in biology, for reviewing an earlier draft of this book, and pointing out sections where changes and summaries would be helpful to readers.

My daughter Jeanne Marie Hodges, son-in-law John Hodges, sister Patricia Lang Townsend, and my son's wife Kelly Renee Lang, were all helpful with their feedback, although they lacked a background in science.

Dr. Stuart Smith, a friend and specialist in marine geology, provided the greatest technical help, next to my son, in reviewing this book. He pointed out several areas that needed more explanation, asked knowledgeable questions, and helped me find better ways to describe certain topics.

In regard to the theory published in GED, and reproduced in the appendices, I received the greatest support from Professor Cynthia Kolb Whitney, the Editor in Chief of *Galilean Electrodynamics* (GED). I made a special trip with my wife to meet her in 1998 to discuss the possible publication of my recently completed manuscript covering this new theory. She suggested physics papers to read, physicists to contact, later approved this theory for publication, and provided additional support that is further acknowledged in the appendices. Professor Whitney is a strong supporter of new ideas in physics, and has published papers in GED and other journals on her own new theoretical concepts in physics and interpretations of experimental results.

I also want to acknowledge the help from the anonymous technical reviewer of my three-part paper in GED. This reviewer provided many technical suggestions and wording changes; these suggestions were unusual because, judging from his comments, the reviewer had an alternative physics theory that is very different from mine, and yet was willing to help.

Of the people who have read and responded to my GED paper, the late Dr. Thomas Van Flandern, an astronomer and specialist in celestial dynamics provided the most technical support for this new theory. He correctly recommended that I more strongly emphasize the wave properties of photons, and that I clarify several phrases and definitions. He also offered suggestions for refining this theory, although it differed from his own unusually creative approach to physics. During our brief correspondence prior to his death in 2009, he expressed a belief that the universe is infinite, which caused me to more-seriously consider an infinite universe.

I want to thank Dr. Thomas E. Phipps, Jr., a physicist whom I have known for about 50 years, for his general support after first hearing of this theory. Although this theory lies outside of modern physics, and his background, he reviewed a recent version of this book and continues with his general support. He has written numerous papers and books on new ideas in theoretical physics.

I also want to thank Professor Everett J. Post, a physicist, for several meetings and correspondence where he kindly provided support and offered suggestions. Although his knowledge of theoretical physics vastly exceeded mine, Dr. Post had the patience to explain important aspects of physics, including his alternative viewpoints in quantum mechanics.

Thanks are also due Dr. Milo Wolff, a physicist with whom I corresponded in 2000-2002. He had developed a new physics theory that involved electrons that pulse, but pulse in a very different manner from my theory. At that time, Dr. Wolff correctly pointed out that my *orbiting electrons* would not appear as cloudy spherical shells, as found in experiments. As a consequence of his comment, I soon noticed that my electron rings are really tiny gyroscopes that will *always turn left* when orbiting; therefore, an orbiting and pulsing electron ring in my theory will never sequentially follow the same path, and will indeed appear as a *cloudy spherical shell.* This finding was made in time for inclusion in Part III of my GED paper at the very end of Section 14.

I also want to acknowledge the help of Dr. Joan Stachiw, a historian, and wife of a late good friend Dr. Jerry D. Stachiw who was the world's leading authority on underwater acrylic windows and spheres. Joan remarkably read and understood an early version of this book in only two hours, and then made suggestions that helped to improve its readability and understanding.

Last, but not least, I want to thank five other people for their help after reviewing different versions of this book. Dr. Timothy Hall, a psychologist and cousin of my wife, provided suggestions that improved the book. Dr.

Judith Golub and her husband Dr. Sidney Golub, who I met on a vacation trip, made helpful suggestions including the need to insert more information on quantum phenomena. I thank Dr. Joan Landsberg, a chemist and friend of my wife's, for suggesting numerous changes in wording and technical description. I thank Robyn Ledwith Mar, an English major, and classmate in high school physics who reviewed a recent version this book that led to further improvements in wording and presentation.

Although I cannot say that any of these people fully understand and agree with this new theory, all of them were helpful and supportive, for which I am most grateful. Only time, future experiments, in-depth analyses, and a professional consensus can determine whether or not this new theory is correct.

14.1 Acknowledgment of Intuition

Because of the key role that intuition played throughout the development of this new physics theory, I felt compelled to acknowledge it. Additionally, I wanted to share some ideas that resulted from decades of thinking about intuition because, contrary to popular belief, I believe that intuition is not mysterious, and instead is something that can be improved, taught, and learned. So please consider or use the following ideas only if they seem reasonable to you.

Intuition and the Mind. About six years ago, the thought suddenly came to mind that *intuition is an instant computer*. This thought seemed so obvious at the time that I wondered why I had not thought of it or read about it before. In any case, this thought led to the idea that the mind primarily consists of an instant computer and a thinking computer. It is well known that the brain is divided into a left brain and a right brain, and that most people believe that the left brain is used for logic and analysis while the right brain is used for creativity. Therefore, for discussion purposes, I will assume that the thinking computer lies in the left brain, and the instant computer lies in the right brain.

Thinking Computer. The idea of a thinking computer is not new; however, no one knows how to design such a computer. In any case, the thinking computer must have all of the hardware and software needed for conscious thought, such as making associations, detecting similarities, using logic, drawing conclusions, solving problems, planning, and creating ideas. A thinking computer must also include a memory bank. In other words, the thinking computer, as envisioned, has what is needed to provide *intelligence*.

Instant Computer. My concept of an *instant computer* is very different. This computer provides intuition, and contains all of the hardware and software needed to provide the *best possible instant answer* to any question or problem. The instant computer must have access to the thinking computer, and especially its memory bank that includes all of the information and conclusions that result from thinking. The instant computer is able to instantly (within a fraction of a second) find new associations, combine ideas, create ideas, summarize, and make decisions. However, no one knows how to design such a computer.

Good and Bad Intuition. One might well ask "Why do some people have very good intuition, while others have very poor intuition?" A short answer is that the minds of people with bad intuitions tend to be filled with conflicting and incorrect information. Consequently, the *best possible answer* that their instant computer can provide is a poor answer. Alternatively, the minds of people with unusually good intuition are filled with well-correlated and correct information. Typically, these people think a lot, use good thought processes, are curious, and learn from their mistakes. In other words, people with unusually good intuition *use both kinds of computers very well*, and have well-organized memories.

Event Memory. For background, and before discussing intuition, I want to describe what I consider to be the two very different kinds of memory that people have: *event memory and thought memory.* Event memory results from experiencing an event. I believe that the mind records events using all five senses; these recordings resemble the two-sense (sight and sound) recordings made by video cameras. The mind titles and subtitles an event by mentally associating the event with a topic, people, places, dates, etc., much like writing a title and subtitles on a DVD.

Event memories must consume a huge amount of memory space, as video photographers well know; and this is why I believe that humans have a short-term memory. Our short-term memory is said to record an entire event, and then permits us to extract those parts that we want to permanently remember, much like editing a video. Although human memory seems to be infinite, I believe that it is important to limit event memory wisely, and not try to remember everything about everything (unless one's profession depends upon such a memory). My reason for this conclusion is that other mental abilities, such as recall, intuition and thought, might otherwise be adversely affected, especially in the long run.

Thought Memory. This second kind of memory is very different from event memory. A thought memory results from mentally extracting ideas and conclusions by using conscious thought. A thought memory summarizes how and why something behaves in different situations. A thought memory results from *understanding* something, in contrast to an event memory, which is a *rote memory*. For best recall, I believe that thought memories should be widely associated with each other. Since thought memories use relatively little memory space, I believe that one should *think more*, not less.

Improve Intuition. Intuition can be improved by learning to use each kind of computer better, and by developing the mental habits associated with people who have good intuition. Note that the instant computer relies on the thinking computer to organize information in the memory so that it can more easily find the best possible instant answer to any question or problem. Also note that, whenever the instant computer finds an instant answer, the thinking computer can then be used to more thoroughly analyze and explore that answer. Alternatively, if the thinking computer is suddenly halted by an unforeseen problem, then the instant computer is often able to immediately solve that problem. *Each computer helps the other.* In other words, the better we learn to use one computer, the better we are able to use the other computer. This win-win situation can lead to a great improvement in both intuition and conscious thought.

Track Back. I firmly believe that every intuitive idea results from a logical, subconscious, very rapid thought sequence. Therefore, one of the first steps in improving intuition is to understand intuition better by tracking back intuitive ideas. Begin by tracking back your simplest intuitive thoughts. Then, continue by tracking back your progressively more-complex intuitive thoughts. Tracking back very complex intuitive ideas is doable, but is probably not worth the effort because it can take hours or days if an intuition is based on a very complex thought sequence.

Organize Memory. Human memory is like a gigantic library. Because of its huge size, it is important to organize one's memory so that information can be recalled when needed. I believe that memory is organized by associating information. There seems to be almost no limit to the number or complexity of associations that can be remembered. The number of associations accumulated in a lifetime must be immense.

Associations are discussed next, and are made as fast as one can think which, for example, is far faster than filling out a library file card. The mind is a wonderful thing!

Associate. Many kinds of information can be associated, such as words, ideas, concepts, devices, methods, procedures, documents, plans, and almost anything. Furthermore, associations can be grouped, and even these groups can be grouped. One might ask: "What are the best associations to make? My answer is: "The best associations are those that work best for you". The reason is that each person's mind and goals are unique. However, I can provide three suggestions that may be helpful. First, it is best to associate information in the way that you want to recall it. Second, it is useful to associate information by summarizing it. Third, unusual associations are important if greater creativity is desired.

Understand. We say that we *understand something* after we learn how *that something* acts in many different situations. This definition applies to objects, ideas, words, and people. I visualize *understanding* as being an island of associated information. With greater understanding, that island grows. Given time, an island will become associated with many other islands of understanding. Given a lifetime, a person's mind will become filled with islands of understanding. If these islands are well interconnected, *then that person is said to be wise*.

Remove Clutter. A quick way to improve intuition is to clear the mind of conflicting information and other kinds of clutter. Begin by analyzing the simplest conflicts in your mind. Any conflict that has an obvious solution disappears by simply associating it with that solution. Similarly, an obsolete conflict disappears the moment that it is associated with the reason for being obsolete. Considerable thought and new information may be required to resolve most of the conflicts in the mind. However, solving conflicts gets easier with experience, and can become almost automatic.

Weaken Negative Events. Negative events such as negative thoughts, habits, and memories can sidetrack or block intuition. It is helpful to eliminate, bypass, or weaken negative events. For example, one can substitute a better thought, habit or memory. Alternatively, one can blunt the effects of negative events by finding the positive lessons learned from each such negative event.

Summarize. I believe that summaries are especially useful; summaries prepared for the instant computer should be general and brief. Alternatively, summaries made for use by the thinking computer can be very detailed and complex. Either computer can handle a wide variety of summaries, including summaries of summaries. Note that summaries include generalizations, similarities, and basic truths.

Think. Conscious thought helps to generalize ideas, summarize and categorize ideas, find associations, develop analogies, create new ideas, organize memory and solve problems. Thinking takes time, but its power is almost unlimited.

Problem Solving. Learning to solve problems better and better will make the process automatic and intuitive. Note that problem statements typically include: (1) a specific goal, (2) conditions that must be satisfied, and (3) things to optimize. A well-worded problem statement is a major step in finding its solution. I firmly believe that a best solution exists for any problem at any given time. A good way to jump-start creativity in problem solving is to first think of all possible solutions, and then conduct a literature search rather than vice-versa. A helpful way to solve very complex problems is to first solve related and simpler problems.

Practice The Thought Processes. Continue to practice and use the various thought processes until they become intuitive. These processes include making associations, finding similarities, making summaries, finding alternatives, analyzing things, and problem solving. Practicing the thought processes is much like practicing riding a bicycle; eventually, they become automatic.

Creative Intuition. Creativity in intuition can be improved in many ways. Most important is to improve all of the thought processes because I believe that they all are involved in optimum creativity. Finding analogies and similarities assists creativity. Conducting leading edge research helps creativity by learning how something behaves in many different situations, and how changes in *that something* affect the results. Research can be done experimentally, analytically, or by doing both. I prefer doing both.

Improve Memory. In theory, a lasting memory is made instantly by associating something with things that you already know. However, recall tends to reduce with time; therefore, memories are strengthened by use. Also, memories can be greatly strengthened by adding parallel associations. For example, to better remember a person's name, make parallel associations between that person's face and name with others who look similar, and who have similar hobbies, backgrounds, ideas, friends, etc. Additionally, a very effective way to remember to do something in the future is to use *visualization*. For example, visualize yourself returning from a trip, entering your house, and then doing *that something*.

Listen With Understanding. If you want to listen, then listen with understanding by associating what you hear with what you know. Also, when listening, avoid debates; I believe that debates too often lead to a battle of wits or ideologies that leads nowhere. Instead, try to work together with the speaker, if possible, to build ideas.

Learn From Mistakes. Ideally, learn from the mistakes of others. Always learn from your own mistakes, and then generalize what you have learned. For example, if something was lost, then determine the cause, think of how to avoid losing that object again, and then think of ways to avoid losing any object.

The Controller of the mind. People have wondered where the controller of the mind is located. During one of my many reviews of this section on intuition, the following intuitive idea suddenly came to mind: "The instant computer is the controller of the mind." Now, I wonder where that idea came from?

Summary. Summarizing, I believe that intuition can be improved, taught, and learned. I have offered a few suggestions on how to do these things, and in the process, I hope to have both *demystified intuition*, and provided some useful ideas.

Intuition Applied in this Book. I used intuition in developing virtually all of the forty new aspects of this physics theory that are listed in Section 13.1. These aspects include ideas that relate to: a new type of spatial fluid that can transform into energy and matter; the use of fluid-dynamic theory to help understand and unify physics; a new pulsing property of photons that explains quantum phenomena; a new description of photons, their creation, physical properties, and movement; the idea of fluid rings, their properties, dynamics, and combinations; gravity as being a pressure force that is an inherent property of all photons and matter; an explanation of the strong and weak nuclear forces, and their properties; and a physical description of proton-neutron pairs. Each of these ideas and concepts were blended together into a new unified theory of physics that has hopefully demystified modern physics, and provided a new physical understanding of physics.

APPENDIX I. PROPOSED UNIFIED FIELD THEORY - PART I: SPATIAL FLUID, PHOTONS, ELECTRONS

Submitted 22 March 1999 by Thomas G. Lang. Reproduced from Galilean Electrodynamics, 11, 43-48 (2000), with permission from the publisher.

This is the first in a series of papers whose primary objective is to provide a new physical theory that offers a deeper understanding for the underlying mechanisms in physics. In this proposed theory, space is hypothesized to consist of a variable-density fluid having special properties that can account for the origin, geometry, characteristics and interrelationships of particles. Here, photons and electrons are developed from spatial fluid with the help of classical hydrodynamic and physical theory.

1. Preface

It is surprising that physicists have made such great progress in developing theories that agree with experiment without being able to physically describe the basic mechanisms. Current theory does not physically describe gravitational, electric or magnetic fields, or explain how gravity is related to electromagnetism. Nor does it explain how photons travel through space and why they act like both particles and waves. Current theory does not describe the structure of electrons, protons, and neutrons, nor does it explain what energy is, what electricity is, and how matter and energy are physically related. The world of quantum mechanics is not physically understood. The following quotations from Nobel Laureate Richard P. Feynman [1] illustrate this lack of physical understanding.

> "*The mechanical rules of 'inertia' and 'forces' are wrong--Newton's laws are wrong--in the world of atoms"..."Here things behave like nothing we know of, so that it is impossible to describe this behavior in any other than analytical ways.*"

> *"It is important to realize that in physics today, we have no knowledge of what energy is."*

> *"There is no explanation of gravitation in terms of other forces at the present time*."

In discussing the wave-particle duality of photons, electrons, protons and neutrons, Feynman [2] stated:

> "*We choose to examine a phenomenon which is impossible, absolutely impossible, to explain in any classical way, and which has in it the heart of quantum mechanics*."

Clearly there is motivation here for developing new theories. The present work proposes a new theory based on hydrodynamics. This first paper describes the theory of spatial fluid as it applies to photons and electrons.

Although space is considered empty, it is difficult to believe that electric fields, magnetic fields, electromagnetic waves, and gravitational fields exist in an empty vacuum. In the proposed theory, space is postulated to be a fluid somewhat like a classical compressible fluid, but with unusual properties. Although it cannot be seen or felt, the spatial fluid is hypothesized to flow, contain vortices, have density and support pressure changes. Unlike any known fluid, it is hypothesized to be the source of photons. This fluid is much different from that proposed in the ether theory of the nineteenth century.

Furthermore, the spatial fluid represents an absolute space, contrary to Einstein's Principle of Relativity.

2. Photons and Time

Light rays are made of photons, as are radio waves, x-rays, gamma rays and other kinds of electromagnetic radiation. The difference between these rays is the energy of their photons. A photon travels at the speed of light c, and consists of a bundle of pure energy at emission and at absorption. How it behaves during travel is not understood, but it is known to act like both a particle and a wave. The energy of a photon is proportional to its frequency, ν_{ph}, with the multiplier Plank's Constant, h:

$$E_{ph} = h\nu_{ph} \qquad (1)$$

A photon is herein hypothesized as a spherical disturbance in space that travels at the local speed of light, transforms smoothly and cyclically between particle and wave states, and spins one way or the other about its travel axis. The spin of photons is hypothesized to be equally distributed each way.

In its particle state, a photon is considered a sphere of compressed spatial fluid that, when growing, draws in fluid from the surrounding space. When shrinking, this fluid is returned. The growth phase resembles a fluid sink in the spatial fluid; and the shrinking phase resembles a fluid source. Nothing is created or destroyed because the sum of spatial mass and photon particle mass remains constant. As the hypothesized photon grows while traveling at light speed c, energy transforms smoothly between kinetic particle energy and wave energy, as shown in Fig. 1.

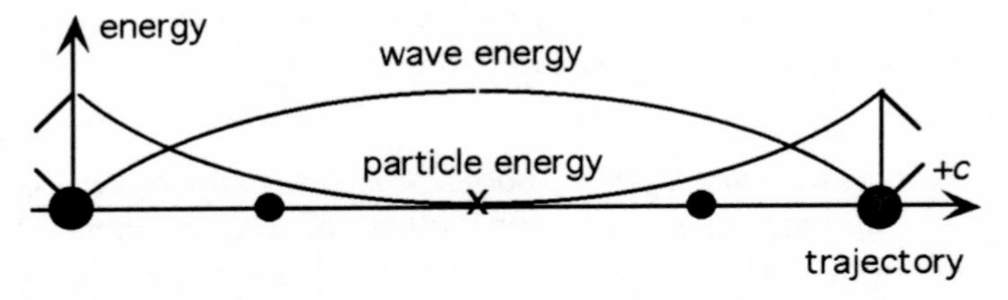

Figure 1. Changes along photon trajectory during one cycle. Energy switches between particle and wave as photon size varies.

The wave portion of photon energy is considered a spherical wave in space that moves at the local speed of light, and extends to infinity as indicated in Fig. 2. Since it must occur during one photon period, fluid mass transfer must take place at near-infinite speed. On the average, a region of reduced spatial density precisely accompanies a photon as it travels through space at speed c.

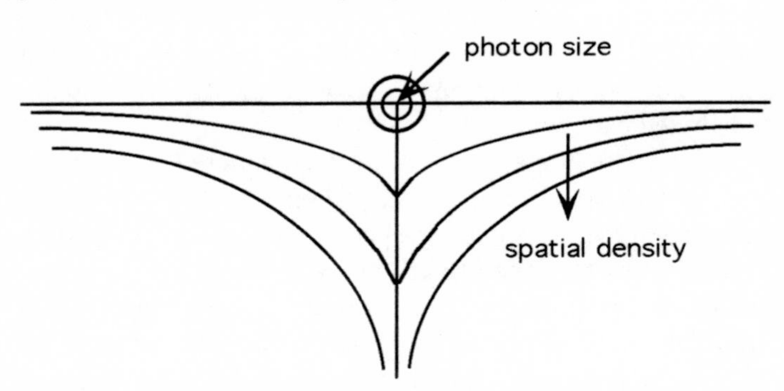

Figure 2. The photon consists of compressed spatial fluid. As it grows, the spatial density of fluid reduces, and *vice versa*.

The photon cycle is envisioned as a dynamic process, somewhat like the repeated cycles of expansion and compression of an underwater explosion bubble or cavitation bubble. A photon particle grows as fluid is compressed by momentum of the incoming fluid. Eventually, particle pressure overwhelms the incoming momentum, and reverses the flow. The photon then loses mass as the outward flow continues. This outwardly flowing fluid flow eventually over-expands causing a negative pressure around the photon that reverses the flow and starts a new cycle. Because the dynamic forces are greater for the more energetic photons, it is now more understandable why higher-energy photons have shorter periods than lower-energy photons.

Although photons have no rest mass, the hypothesized photon particles have a dynamic mass that cyclically varies from zero to a maximum, and back again. When a photon particle reaches its maximum physical mass, M_{ph}, its energy is totally in the form of kinetic energy, so:

$$E_{ph} = \frac{1}{2} M_{ph} c^2 \tag{2}$$

Defining m_{ph} as the average mass of a photon particle at the point in the cycle when photon energy is equally divided between particle energy and wave energy, then:

$$\frac{1}{2} E_{ph} = \frac{1}{2} m_{ph} c^2 \text{ or } E_{ph} = m_{ph} c^2 \tag{3}$$

By definition, photon wavelength is photon period, T_{ph}, multiplied by c, which provides the following relationships when combined with (1) and (3):

$$\lambda_{ph} = T_{ph} c = c / \nu_{ph} = ch / E_{ph} = h / m_{ph} c \tag{4}$$

It is known that photons are created as the result of different kinds of energy exchanges. Whatever the energy exchange, it is reasonable to assume that energy, linear momentum and angular momentum are conserved. For example, if only one photon is generated during an energy exchange, then angular and linear momentum must be imparted to other particles. Any number of photons can be generated. If energy converts into spatial fluid, and this spatial fluid converts into photon energy, then a minimum of two pairs of photons must be generated, as shown in Fig. 3a in order to conserve momentum and satisfy spin symmetry. This is so because two opposite-moving photons must have the same spin in order to conserve angular momentum. Therefore, another pair of photons must be created to equalize spin in each direction. If two such photons are created along the same path, then each photon would have a companion with opposing spin, as shown in Fig. 3b.

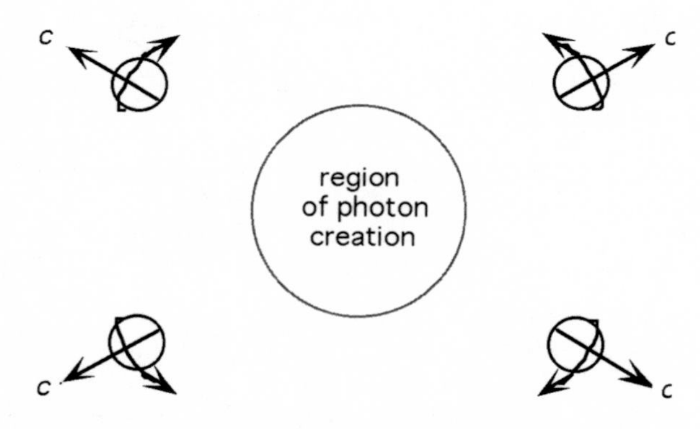

Figure 3a. Creation of two pairs of photons.

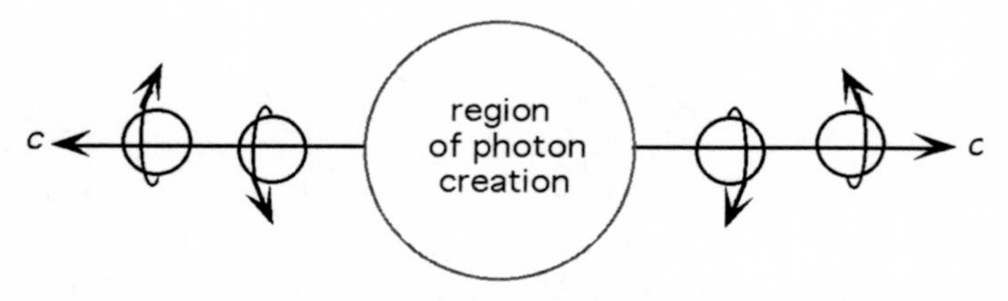

Figure 3b. Special case where paths are identical.

In summary, a hypothesized photon is a spherical disturbance in space that travels at the local speed of light, cyclically transforms smoothly between particle and wave states at a frequency proportional to its energy, spins one way or the other about its travel axis, is accompanied by a region of reduced spatial density, and in its particle state comprises a sphere of compressed spatial fluid.

The time taken by a photon to travel a given distance is simply that distance divided by the speed of light. Consequently, if the speed of light were zero, then an infinite time would be needed for light to travel any given distance; no action would take place, and there would be no meaning to time. On the other hand, if the speed of light were infinite, then all action would occur simultaneously; and time would again be meaningless. Therefore, the existence of a finite speed of light permits sequencing of events which leads to the idea of time.

3. Electrons and Positrons

An electron is known to be the lightest stable form of matter, and is identical to a positron except for electric charge and spin direction. It is further known that if an electron and a positron are brought together at rest, then they annihilate into two gamma ray photons. It is conjectured that the inverse occurs whereby electrons are formed from gamma photons.

If two of the hypothesized photons have counter-clockwise spin, the same phase, a specific energy level, and approach each other on a near-collision course, then they will circle and orbit, as shown in Fig. 4. The required attraction force is the known hydrodynamic attraction force between two sources or two sinks. Because the photons are in phase, they will both act as sources or sinks at any instant. While circling each other, it is hypothesized that each photon incrementally elongates

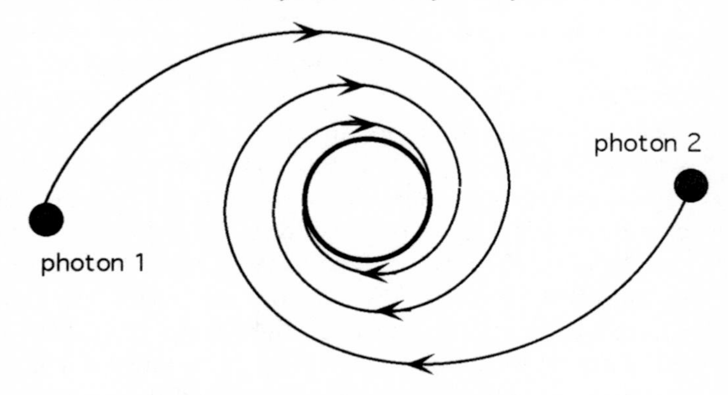

Figure 4. Electron formation.

along its circular path, a little more each cycle, until they merge along their orbital paths. The result is a ring of pulsating, spinning, compressed spatial fluid, as shown in Fig. 5. This ring is considered an electron.

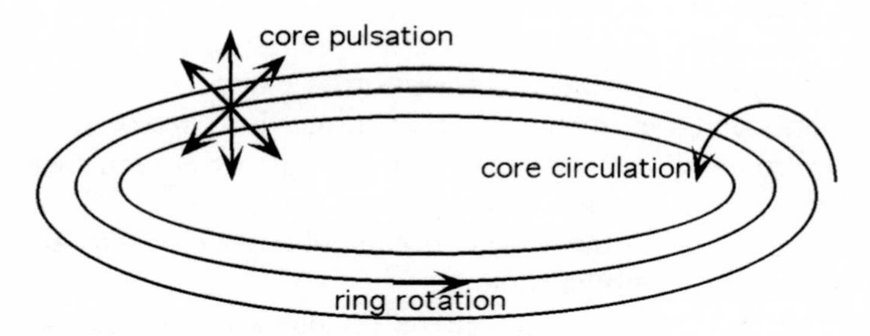

Figure 5. Hypothesized electron.

Because the electron ring contains twice the energy of a single photon, and frequency is proportional to energy, its frequency is about twice that of its two initial photons. Binding energy accounts for any small difference. For reasons analogous to Beckmann's stability analysis for an electron orbiting a nucleus in an atom [3], an electron ring rotates once per cycle for dynamic stability.

In summary, a hypothesized electron (Fig. 5) smoothly and cyclically transforms between particle and wave states, is accompanied by a region of reduced spatial density, and in its particle state appears as a ring of compressed spatial fluid that rotates at c and has a core that spins counter clockwise while growing from almost nothing to a maximum and back again.

The hypothesized electron cycle must be purely elastic since there is no known energy loss with time. Therefore, the energy of such an electron is constant, and always equals the sum of its kinetic particle energy and wave energy. A positron is considered identical to an electron, except its core spins clockwise. Because photon spin is hypothesized to be equally distributed, it is reasonable to assume that the same number of electrons and positrons are created.

The hypothesized electrons are intimately connected with the surrounding spatial fluid, and induce two kinds of vortices in the spatial fluid. One kind of vortex is induced by rotation of the ring, as shown in Fig. 6.

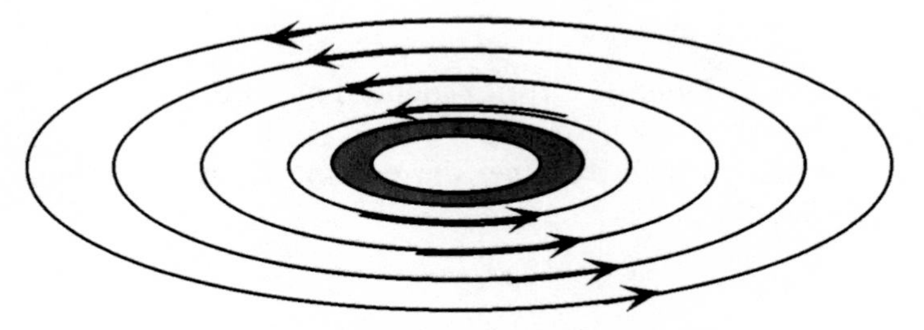

Figure 6. Vortex induced by rotation.

The other kind of vortex is a circular vortex induced by circulation of the core of the ring, as shown in Fig. 7, and behaves like a smoke ring in air. The circular vortex produces a flow of spatial fluid through the center of the ring, and a reverse flow around the outside of the ring. If an electron is free to move, the circular vortex induces a velocity upon itself, like a smoke ring, causing the electron to move in the direction of flow induced through its center at a fixed, natural speed. Electron ring rotation is counter clockwise when viewed in the direction of travel.

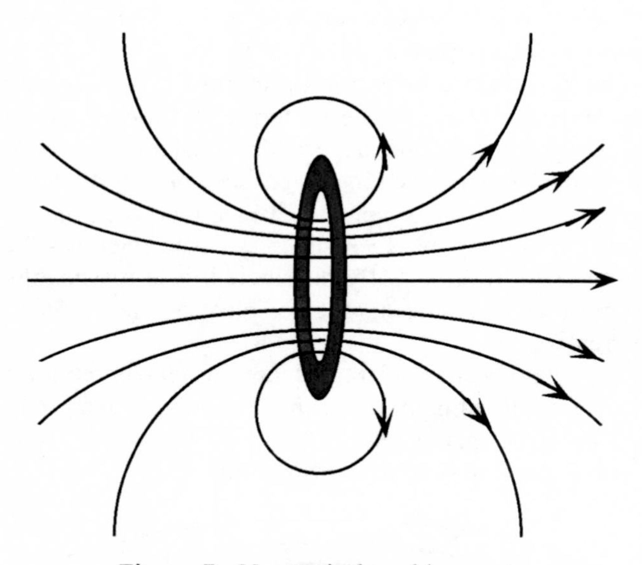

Figure 7. Vortex induced by core.

At the particle extreme, the energy of an electron is fully kinetic, and is:

$$E_e = \frac{1}{2} M_e c^2 \quad (5)$$

where M_e is the maximum particle mass of an electron. If m_e is electron particle mass when half of its energy is kinetic, and the other half is wave energy, then the "particle half" of the total electron energy is:

$$\frac{1}{2} E_e = \frac{1}{2} m_e c^2 \text{ or, } E_e = m_e c^2 = \text{total energy} \quad (6)$$

This equation is the well-known Einstein relationship, but derived much differently.

Assuming that an electron rotates once per cycle, expressions for its circumference, radius, mass and energy are developed with the help of (1) and (6):

$$\lambda_e = 2\pi R_e = c/\nu_e = ch/E_e \tag{7}$$

rearranging,

$$R_e = c/2\pi\nu_e = ch/2\pi E_e \tag{8}$$

$$m_e = h/2\pi R_e c \tag{9}$$

and

$$E_e = h\nu_e = h/T_e = ch/\lambda_e = ch/2\pi R_e \tag{10}$$

The proposed electron ring model supports the ring-shaped electron model proposed by Bergman and Wesley [4] whose derivation was based on electromagnetic theory. However, the model presented herein is quite different because it includes spatial fluid, spin of the electron core, pulsation of the core, transformation between wave and particle modes, and is hypothesized to be created from photons.

4. Various Results

Electron natural speed. A fluid ring whose core spins about the core axis is called a vortex ring. The hypothesized electron is a type of vortex ring. A smoke ring is another type of vortex ring. All vortex rings have a natural speed, which is the speed induced upon itself by rotation of its core. Let Γ_{2dce} be the average 2-dimensional circulation (*i.e.*, core speed times circumference) of an electron core, R_e be the ring radius, and R_{ce} be the average core radius. Substituting these expressions into the equation presented by Milne-

Thompson [Ref. 5; p. 562] for vortex ring speed, the natural speed of the hypothesized electron (such as in a hydrogen atom) is:

$$V_e = \frac{\Gamma_{2dce}}{4\pi R_e}\left[\ln(8R_e / R_{ce}) - 1/4\right] \quad (11)$$

Angular momentum and magnetic moment of electrons. A free electron has both angular momentum and magnetic moment. The proposed electron model provides values of angular momentum and magnetic moment in agreement with experiment. The angular momentum of a hypothesized electron ring is:

$$J_e = K_e m_e c R_e \quad (12)$$

where K_e is the ratio of average electron kinetic energy to total energy. As discussed in the section "Electrons and positrons", this ratio is 1/2. Substituting $K_e = 1/2$, and (9), into (12):

$$J_e = \tfrac{1}{2}\left(h / 2\pi R_e c\right) c R_e = h / 4\pi = \hbar / 2 \quad (13)$$

"Spin" is defined as $S = J / (h / 2\pi)$. Therefore, the spin of the hypothesized electron is $S = \pm 1/2$, where the sign depends upon its orientation.

If it is assumed that the so-called charge of an electron resides in the ring of a hypothesized electron, then its magnetic moment, μ_e, is "charge" times frequency times the enclosed area:

$$\mu_e = -q_e \left(c / 2\pi R_e\right) \pi R_e^2 = -\tfrac{1}{2} q_e c R_e$$

Substituting (8):

$$\mu_e = -\tfrac{1}{2} q_e c\left(h / 2\pi m_e c\right) = -\left(q_e / m_e\right)\left(\hbar / 2\right) \qquad (14)$$

Both J_e and μ_e are in agreement with experiment.

Equivalence of mass and energy. Photons are considered to be pure energy, and have no rest mass because they cannot be at rest. Electrons have rest mass because they can remain at rest although they are hypothesized to be in extreme motion. The proposed theory physically explains how energy, in the form of photons, converts into electrons which have mass.

Alternatively, it is known that mass converts into energy. The proposed theory helps to physically explain how this conversion occurs. Assume that an electron and a positron came together at rest, causing the rings to break into two components. To satisfy conservation of energy and momentum, each component must travel at c in opposite directions. It is conjectured that, when two such rings come together, their orbital attraction forces are cancelled, causing each ring to straighten into a spinning core that moves endwise at c. Each straight-line core would continue to cycle between particle and wave states. It is further conjectured that each core shortens under hydrodynamic forces, a little more each cycle, until each becomes a spinning, pulsating sphere, herein defined as a photon.

Equivalence of mass, energy and space. The hypothesized formation of photons from spatial fluid, cyclical conversion between particle and wave states, formation of electrons and positrons from photons, and postulated conversion of electrons back into photons, illustrate equivalence between mass, energy and spatial fluid.

Missing, is the steady, non-cyclical conversion of photons back into spatial fluid. To complete the symmetry, it is hypothesized that photon movement through space is not frictionless, and that a very small amount of energy is lost during each cycle as it travels through space. This lost energy converts into an increase in spatial density. In other words, a photon slowly converts back into spatial fluid as it travels through space, and ultimately disappears. Nothing is created or destroyed.

Red shift. A consequence of the above hypothesis is that the red shift in light received from other galaxies is at least partly caused by photon energy loss, and not completely caused by a Doppler shift. In the extreme case, the red shift is fully caused by energy loss, thereby implying that the universe is infinite, and not expanding from the big bang. The so-called 'tired light' theory has been proposed many times, but was discarded because there was no place for the lost energy to go. The proposed theory provides that place, assuming that photon cycles are not fully elastic.

Static Force on electrons. Hydrodynamic theory shows that a side force is generated on a rotating cylinder when placed in a fluid flow. This force is perpendicular to the fluid velocity, V, and is directed toward that side of the cylinder that rotates in the direction of the fluid flow, as shown in Fig. 8.

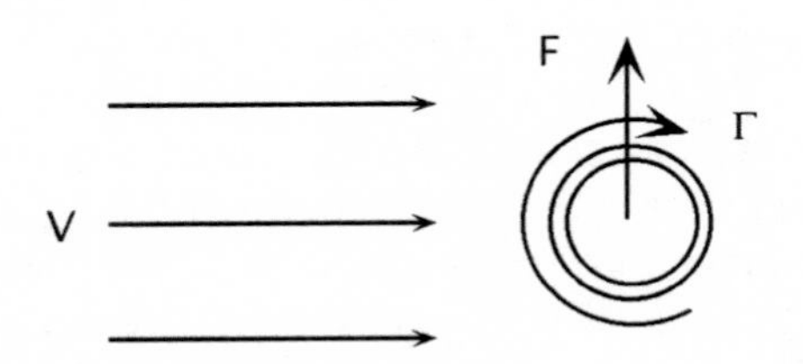

Figure 8. Side force on a rotating cylinder.

The force per unit length is $F = \rho V \Gamma$ where ρ is fluid density and Γ is the circulation. A similar side force is generated on a rotating ring, spinning golf ball or spinning baseball. The force on a hypothesized electron ring is:

$$F_e = \rho V_f \Gamma_e \tag{15}$$

where V_f is the relative velocity of the spatial fluid in the plane of the ring, and Γ_e is the three-dimensional circulation generated by an electron ring.

Electron orientation. If an electron moves through the spatial fluid, and its plane is angled to the flow, as shown in Fig. 9, then equal and opposite forces are generated on its spinning core (see Fig. 8). Although the total force is zero, the moment produced by these forces causes the electron to point into the flow. An electron is defined to 'point' in the direction of the self-induced flow through its center.

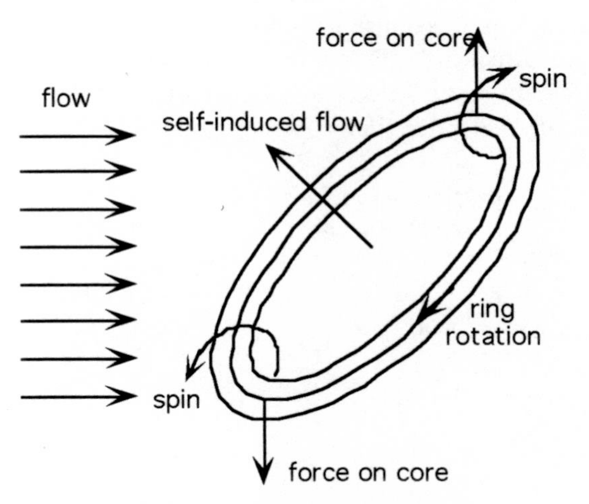

Figure 9. Electron dynamics.

Opposing force between electrons. In view of the previous section, two isolated electrons, if pinned in space, will rotate under the influence of the resulting induced velocities until they point in the same direction, thereby lying in the same plane. If separated by distance r, classical hydrodynamics shows that the rotation of their rings induce the following velocity on each other, perpendicular to a line joining their centers:

$$V_{ie} = \Gamma_e / 4\pi r^2 \qquad (16)$$

These induced velocities generate a force, F_{ee}, on each ring, as shown by substituting (16) into (15) where $V_{ie} = V_f$:

$$F_{ee} = \rho \Gamma_e^2 / 4\pi r^2 \qquad (17)$$

This force is directed away from the other ring (see Fig, 10). Two isolated positrons also generate an opposing force. However, an electron and a positron generate an attraction force because the positron ring rotates in the opposite direction. These forces are considered the so-called electrostatic forces, and the induced fluid flows are the so-called electric fields.

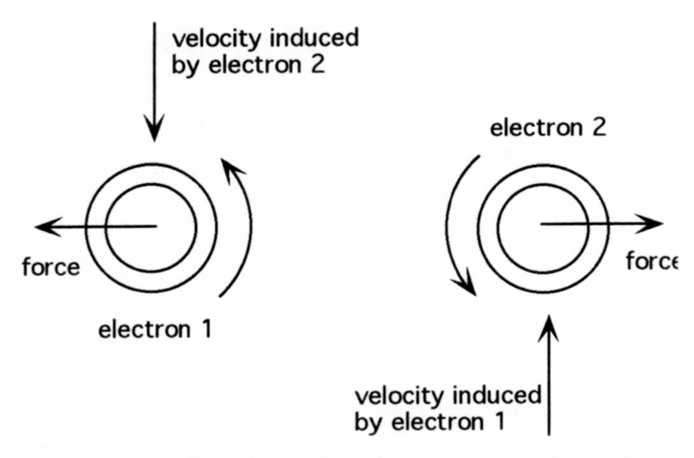

Figure 10. Repulsion force between two electrons.

Equivalence to electric charge. Expressions for circulation of the hypothesized electron in terms of electric charge, and the fine-structure constant, can now be developed. The well-known electron-electron force, where $-q_e$ is the charge on an electron, is:

$$F_{ee} = q_e^2 / 4\pi\varepsilon_0 r^2 \quad (18)$$

Equating (17) and (18):

$$\Gamma_e = q_e \big/ \sqrt{\rho\varepsilon_0} \quad (19)$$

The fine-structure constant is defined as:

$$\alpha = q_e^2 \big/ 4\pi\varepsilon_0 \hbar c = 1/137.036 \quad (20)$$

Combining (19) and (20), and substituting $\hbar = h/2\pi$:

$$\Gamma_e = -\sqrt{2\alpha h c/\rho} \quad (21)$$

Alternative electron-electron interactions. The proposed electron model permits alternative interactions between electrons that help explain their behavior in electron clusters, atoms, *etc.* For example, two nearby electrons with opposite spin will attract, as shown in Fig. 11a. The vortex core of each electron induces a fluid velocity on the adjacent core, generating an attraction force. Two such electrons can be dynamically stable if they orbit a central axis, as shown in the Figure.

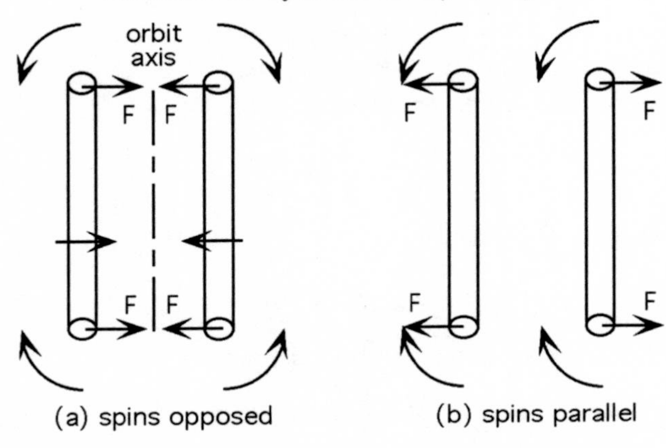

Figure 11. Forces between parallel electron pairs.

If instead, the electrons had parallel spin, as shown in Figure 11b, then they would repel. This latter arrangement illustrates how electrons can repel in a geometry different from that shown in Fig. 10.

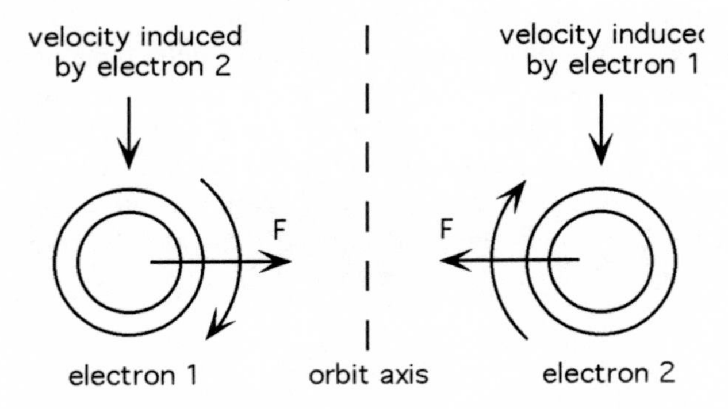

Figure 12. Attractive force between orbiting electrons.

Another example of electron-electron attraction is illustrated in Fig. 12, where two electrons lie in the same plane, but point in opposite directions. The rotation of each electron ring induces a velocity on the other that produces an attraction force. This geometry is dynamically stable if the electrons orbit an axis lying midway between them, as shown in the figure. Several such pairs of electrons could orbit a common axis, forming a stable ring of electrons. Such a ring could either orbit as an independent unit, or could orbit an atomic nucleus. One or more such rings could alternatively orbit a nucleus off center, and be stabilized by the attraction force of nuclear protons. Except for electron movement, the latter case is somewhat analogous to Lucas [6] who hypothesized rings of stationary electrons positioned in planes that lie off center from the nucleus of an atom.

5. Summary and Conclusions

Space is here envisioned to consist of a variable-density fluid. A photon is hypothesized to be a disturbance in space that travels at the local speed of light, cyclically transforms smoothly between particle and wave states at a frequency proportional to its energy, and consists of a sphere of spinning, compressed spatial fluid in its particle state. An electron is hypothesized to result from two orbiting photons, cyclically and smoothly transforms between particle and wave states at a fixed frequency, and consists in its particle state of a ring that rotates at c and whose core is compressed spatial fluid that spins counter clockwise. Positrons are identical to electrons, except for opposite core spin.

The origin, geometry, characteristics and interrelationships of the hypothesized photons, electrons and spatial fluid are developed with the help of classical hydrodynamic and physics theory. Mass, energy and space are hypothesized to be interchangeable. Electron charge is analogous to hydrodynamic circulation of an electron ring. The angular momentum and magnetic moment of the hypothesized electrons agree with experiment.

The following are some of the more unusual conclusions derived from the proposed theory:

- All fields, forces, energy, and matter are hydrodynamic properties of the spatial fluid.
- Mass, energy and space can transform back and forth into each other.
- Photons transform between particle and wave states, and are the source of electrons and positrons.
- Electrons and positrons are energy in the form of matter.
- Electrons are pulsating rings that rotate at light speed with spinning cores of compressed spatial fluid.
- Everything is dynamic, nothing is at rest, and nothing is solid.
- Electrons typically repel, but they can also attract if arranged in special geometries.

Acknowledgements

The author would like to acknowledge the help of Dr. Cynthia Whitney, and the anonymous reviewer, for their many suggestions and thought-provoking comments.

References

[1] R. P. Feynman, R. B. Leighton, and M. Sands, **The Feynman Lectures on Physics**, Volume I, pp. 2-6, 4-2 (Addison-Wesley, New York, 1963).

[2] R. P. Feynman, **Six Easy Pieces**, p. 117 (Addison-Wesley, New York, 1995).

[3] P. Beckmann, **Einstein Plus Two**, pp.124-127, (Golem Press, P.O. Box 1342, Boulder, CO 80306,1987).

[4] D. L. Bergman, and J. P. Wesley, "Spinning Charged Ring Model of Electron Yielding Anomalous Magnetic Moment", Galilean Electrodynamics **1**, 63-67, (1990).

[5] L. M. Milne-Thompson, **Theoretical Hydrodynamics**, fourth edition (Macmillan, New York, 1960).

[6] J. Lucas, "A Physical Model for Atoms and Nuclei", Galilean Electrodynamics 7, 3-12 (1996).

APPENDIX II. PROPOSED UNIFIED FIELD THEORY – PART II: PROTONS, NEUTRONS AND FIELDS

Submitted 29 July 1999 by Thomas G. Lang. Reproduced from Galilean Electrodynamics, 12, 103-107 (2001), with permission from the publisher.

This is the second in a series of papers whose primary objective is a new physical theory that offers a deeper understanding for the underlying mechanisms of physics. In this proposed theory, space consists of a variable-density fluid having special properties that can account for the origin, geometry, characteristics and interrelationships of energy and matter. This Part II includes the formation and characteristics of protons and neutrons, electromagnetic fields, gravitational fields, and other topics including electron dynamics and currents.

6. Introduction

This is the second in a series of papers, so Equations, Figures, and References continue from Part I (Ref. [7] here). In Part I, a photon is hypothesized as a disturbance in space that travels at the local speed of light, cyclically transforms smoothly between particle and wave states at a frequency proportional to its energy, and in its particle state consists of a sphere of spinning, compressed spatial fluid. The size, density and pressure of a hypothesized photon increases as fluid enters from surrounding space. After reaching a maximum size, a photon shrinks to zero as its compressed fluid expands back into the surrounding space. Throughout the Universe, photon spin is divided equally between opposite directions.

An electron is hypothesized to result from two photons that approach each other and orbit, then spread out along their paths to form a single ring that cyclically and smoothly transforms between particle and wave states. In its particle state, an electron is hypothesized to consist of a ring of compressed spatial fluid that rotates at the speed of light and has a core that spins counter-clockwise about its direction of motion. The electron core increases in size, density and pressure as fluid enters from surrounding space. After reaching a maximum size, the core shrinks to zero as its compressed fluid expands back into the surrounding space. Positrons are identical to electrons, except for opposite core spin. The numbers of electrons and positrons generated are equal.

7. Protons and Neutrons

Proton formation. A proton is 1836.15 times more massive than an electron or positron. A proton is hypothesized to be a ring that is geometrically and dynamically similar to a positron, and results from the sequential amalgamation of 918 electrons and 919 positrons. The small difference in mass from the total of 1837 particles is the so-called binding energy of a proton. In Part I, it was hypothesized that the same number of electrons and positrons are formed. Since hypothesized protons have one extra positron, and the number of electrons and protons in the Universe are equal, the question "Where are all the positrons?" is now answered.

The first phase in the hypothesized formation of a proton begins with a single positron that attracts two electrons, one on each side, as shown in Fig. 13a. When the electrons contact the positron, their peripheral speeds mesh. It is conjectured that meshing avoids annihilation. After meshing, the positron pivots 180 degrees, and now rotates in the same direction as the electrons. The three rings then merge, forming a single, new ring. This new ring rotates at the speed of light, is geometrically and dynamically similar to an electron, but has three times its mass and frequency, and one-third its diameter. This new ring is herein called an "e-3" ring.

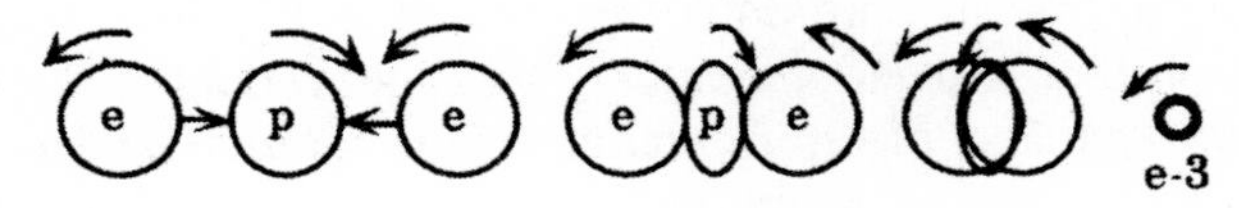

Figure 13a. Hypothesized formation of an e-3 ring, first phase in formation of a proton. A positron attracts two electrons, the rings mesh, the positron pivots 180 degrees to rotate in the same direction as the electrons; the three rings then merge into a single ring with 3 times the mass and 1/3 the radius.

The second phase in the hypothesized formation of a proton, begins with the attraction of two positrons by the e-3 ring, as shown in Fig. 13b. The two positrons mesh with the e-3 ring and dynamically merge, as in Fig. 13a, to form a new ring called a p-5 ring. This p-5 ring rotates at the speed of light, is geometrically and dynamically similar to a positron, and acts like it, but has five times its mass and frequency, and one-fifth its diameter.

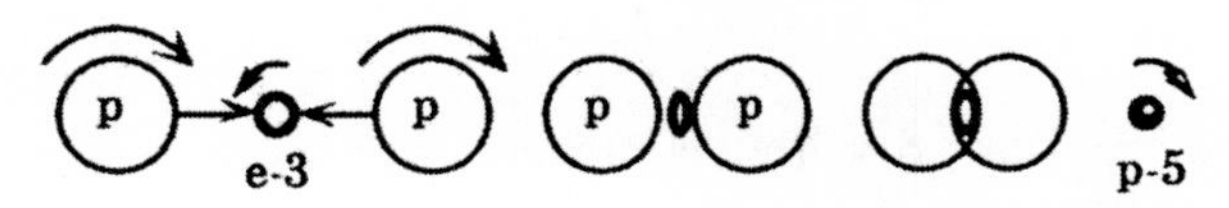

Figure 13b. Hypothesized formation of a p-5 ring, second phase in formation of a proton.

These phases continue, possibly accelerating by merger of larger sub-units, such as two p-3's and one e-3 to form a p-9. Growth then halts after 918 electrons and 919 positrons have merged, forming a proton. It is hypothesized that the merging halts because a proton is physically too small to attract and merge with two more electrons.

Proton mass, radius and density. It is logical to assume that mass, energy, angular momentum, and circulation are conserved during proton formation. The binding energy of a proton, and its equivalent mass, is accounted for by assuming that N electrons and positrons combine to form a proton, so

$$m_p = N m_e \tag{22}$$

Letting V_{rp} be the ring velocity of protons, the energy of N electrons is equated to that of one proton, or $N m_e c^2 = m_p V_{rp}^2$. Substituting (22),

$$V_{rp} = c \tag{23}$$

Therefore, electron and proton rings both rotate at the speed of light. The angular momentum of each sequential particle remains constant, but changes sign. The final result is a proton which has the same angular momentum as a positron, so $m_p R_p^2 \omega_p = -m_e R_e^2 \omega_e$. Since the magnitudes $R\omega_p = R\omega_e = c$, it follows that $m_e R_e = m_p R_p$. Substituting (22),

$$R_e / R_p = N \tag{24}$$

Therefore, the radii of electrons and protons are inversely proportional to their masses.

We now address core density. Let ρ_e and R_{ce}, and ρ_p and R_{cp}, be the maximum densities and core radii of electrons and protons, respectively, then the maximum mass of a proton is N times the electron mass, so $\rho_p(2\pi R_p)(\pi R_{cp}^2) = N\rho_e(2\pi R_e)(\pi R_{ce}^2)$. Substituting (24) and rearranging yields $\rho_p / \rho_e = N^2 (R_{ce} / R_{cp})^2$. It is reasonable to assume that, as for electrons (11), ρ_p varies with the fourth power of energy or mass, so:

$$\rho_p / \rho_e = N^4 \quad (25)$$

$$R_{ce} / R_{cp} = N \quad (26)$$

$$R_{ce} / R_e = R_{cp} / R_p \quad (27)$$

Consequently, hypothesized electrons and protons are geometrically similar. This result supports Bergman [8], who used electromagnetic theory to show that electron and proton rings are geometrically similar. However, the present model is quite different; it includes spatial fluid, spin of the electron core, pulsation of the core, transformation between wave and particle modes, and is hypothesized to be created from electrons and positrons.

Nuclear particles. More nuclear particles are found each year. The hypothesized formation of protons provides a possible explanation for this. Any new particle formed in each of the 918 phases in the hypothesized proton formation might reappear in proton annihilation. So *all* charged particles are hypothesized to be rings rotating at light speed, geometrically and dynamically similar to electrons and positrons. In addition, charged particles can combine, so particle mass can exceed proton mass.

Proton-neutron pairs. A neutron is unstable, and disintegrates in about 10 minutes into a proton, electron and a so-called neutrino. Alternatively, a proton-neutron pair is very stable. It is postulated that a proton-neutron pair physically consists of two protons bound together by an e-3 ring, as shown in Fig. 14 (which is drawn approximately to scale). This postulate is reasonable because a neutron is known to be heavier than a proton by about three electron masses, which is the mass of an e-3 ring.

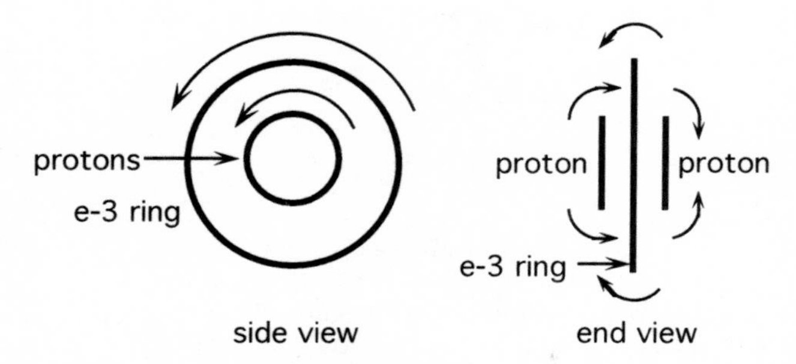

Figure 14. Hypothesized neutron-proton pair. The hydrodynamic forces between the spinning cores of these particles are extremely strong due to their sub-electronic spacing. These forces are hypothesized to be the so-called 'strong force' between nucleons, and are much greater than so-called electrostatic forces.

It is hypothesized that these three rings form a stable unit if they rotate about the same axis in the same direction, with the protons located on opposite sides of the e-3 ring. The fluid velocities induced by adjacent rotating cores of these rings cause the e-3 core to attract the proton cores, while the proton cores repel each other. Because the rings are flexible, and the e-3 ring has much less mass, equilibrium is reached after the e-3 ring greatly contracts, while the proton rings slightly expand, from their free-state radii.

Neutrons. If the hypothesized geometry of a proton-neutron pair is correct, then a neutron must consist of one proton and an e-3 ring, each rotating in the same direction about a common axis, as shown in Fig. 15. Relative to the geometry shown in Fig. 14, the core-core forces are greatly reduced, so the e-3 ring is much larger than that in the proton ring. The resulting geometry would appear to be marginally stable, which explains why a free neutron can remain intact for as long as 15 minutes on average.

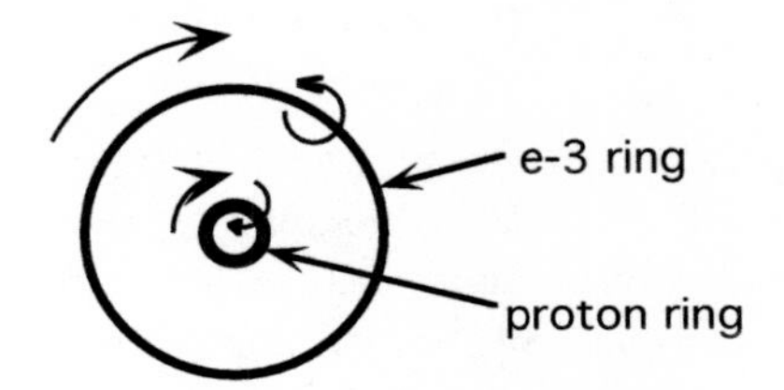

Figure 15. Hypothesized neutron.

Angular momentum and magnetic moments of protons. Since protons and electrons both rotate at light speed, and the product of their mass and radius is the same, their angular momentum must be identical, and therefore equal to (13), so

$$J_e = J_p = \hbar / 2 \tag{28}$$

'Spin' is defined as $S = J / \hbar$. Therefore, the spin of these elementary particles is $S = \pm 1/2$, where the sign depends upon the orientation of the particles.

Analogous to the derivation of the magnetic moment of a hypothesized electron (14), the magnetic moment of a hypothesized proton ring is one nuclear magneton, or

$$\mu_p = \frac{q_e}{m_p}\frac{\hbar}{2} = \frac{q_e}{m_p} J_p \tag{29}$$

Consequently, the magnetic moment of a proton is 1/1836 that of an electron. This expression agrees with that derived by Bergman [8], who states that it agrees with the measured magnetic moment of free protons.

Magnetic moments of protons and neutrons in a nucleus. The respective measured magnetic moments of a proton and a neutron in a nucleus, according to Blatt [9] and others, are:

$$\mu_{pn} = 2.79\frac{q_e}{m_p}\frac{\hbar}{2} \quad \text{and} \quad \mu_{nn} = -1.91\frac{q_e}{m_p}\frac{\hbar}{2} \tag{30}$$

Subtracting the magnetic moment of one nuclear proton from that of a hypothesized nuclear neutron, the magnetic moment of the associated e-3 ring is -1.91-2.79=-4.70 nuclear magnetons. This result means that the radius of a contracted e-3 ring in a neutron is 4.70 times that of a free proton, or 10^{-15}m. The experimental expression for the radius of a nucleus according to Blatt [9] is $R_A \approx 1.5 \times 10^{-15} A^{1/3}$m, where A is the number of nucleons. Although this expression is valid for A>20, and A=1 for a neutron, the radius of a neutron is in the vicinity of the expression.

Equations (30) show that the cores of the three rings in a hypothesized proton-neutron pair lie very close together. The diameter of the nuclear e-3 ring is only 4.70/2.79=1.68 times that of a nuclear proton. Therefore, the distance between the ring cores is less than one nuclear proton diameter, or 1/918 electron radius. Consequently, the hydrodynamic forces between cores is far greater than typical electrostatic forces between electrons and protons, and is here considered to be the "strong force" that binds nucleons.

Elasticity of particle rings. It is now possible to determine the elasticity of particle rings. Continuing from above, the e-3 ring contracted from a diameter N/3 = 612 times that of a free proton, to 1.68×2.79=4.69 that of a free proton, or a contraction ratio of 612/4.69=130.5. Alternatively, the proton expanded 2.79 times from its free state. Therefore, the e-3 ring contracted 130.5/2.79 = 46.8 times more than the proton expanded. Because two protons are combined with one e-3 ring, the force on the e-

3 ring is twice that on the proton ring. If "ring elasticity" is defined as "change in diameter per unit core-core force", and symbolized by ζ, then an e-3 ring has a ζ that is 46.8/2=23.4 times that of a proton ring. This ratio is approximately equal to the square root of m_p / m_{e-3}. This relationship should hold for other particles, including electrons, so a general expression for the ring elasticity of any particle x is

$$\zeta_x \propto \sqrt{m_x} \tag{31}$$

8. Magnetic and Gravitational Fields, and the Michelson-Morley Experiment

Magnetic fields. If an electron or proton is fixed in space, then its spinning core induces a flow of spatial fluid through its ring. This flow produces a return flow around the outside of the ring. Consequently, hypothesized electrons and protons are considered as tiny pumps. If an equal number of electrons and protons are aligned parallel, as in Fig. 16, a flow of spatial fluid is induced through their centers.

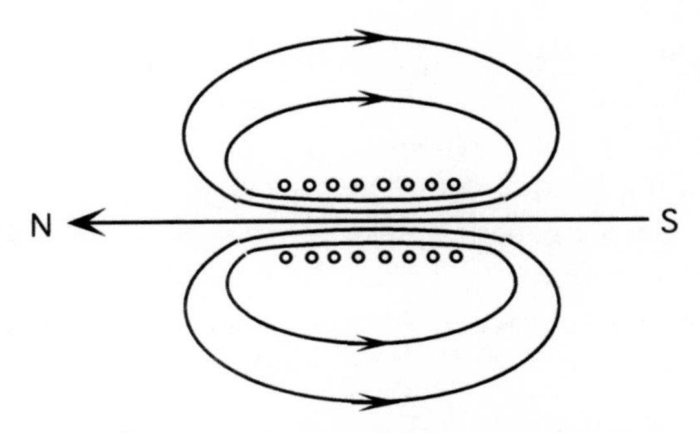

Figure 16. Flow through aligned electrons and protons. The flow pattern remains similar even if the particles are offset from the center line.

The return flow is considered a magnetic field, such as that surrounding a bar magnet, which can be observed by sprinkling iron filings on paper placed above a magnet. The fluid motion induced by electron and proton ring rotation cancels. Therefore, the only hydrodynamic flow is that induced by the pump-like action of aligned electrons and protons.

Experiments show that the magnetic field of a bar magnet does not rotate with the magnet. This result agrees with the proposed model because the flow pattern is fixed, and independent of rotation about the north-south axis.

Gravitational fields. As discussed in Part I, hypothesized photons, electrons and positrons are accompanied by a region of reduced spatial density; therefore, all mass and energy is accompanied by a region of reduced spatial density. It is further hypothesized that:

(a) spatial pressure reduces when spatial density reduces,

(b) a force is exerted on any mass or photon located in a spatial pressure gradient that is equal to the pressure gradient multiplied by its effective volume (*i.e.*, mass divided by the local spatial density). This force, as illustrated in Fig. 17, is the so-called gravitational attraction between masses, and is considered a buoyant force similar to that produced on submerged bodies. Gravity is here considered a "pushing" phenomenon, not the "pulling phenomenon" in current physics. Let P_1 be the reduction in spatial pressure caused by mass #1, and Vol_2 be the effective volume of mass #2. The buoyant force urging the two masses together (*i.e.*, gravitational force) is

$$F_G = \frac{d\Delta P_1}{dr} Vol_2 = \frac{d\Delta P_1}{dr} \frac{m_2}{\rho} \tag{32}$$

The known gravitational force between two masses is

$$F_G = Gm_1m_2 / r^2 \tag{33}$$

where G is the gravitational constant. Equating (32) and (33) and integrating yields the reduction in spatial pressure surrounding any mass m_1 as inversely proportional to distance r.

$$\Delta P_1 = -\rho Gm_1 / r \tag{34}$$

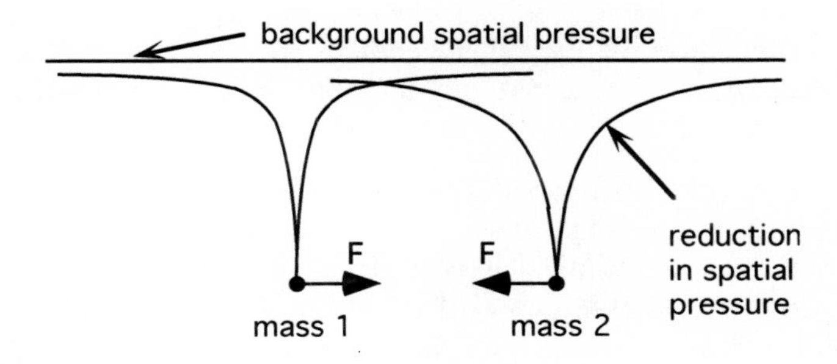

Figure 17. Gravitational pressure field.

Michelson-Morley experiment. The ether theories of the 1800's were abandoned primarily because of the Michelson-Morley experiment [10]. This experiment showed that the speed of light is the same whether measured parallel or perpendicular to the earth's path around the sun. The various ether theories at that time hypothesized that the ether was stationary, and that movement of the earth around the sun would cause a difference in the speed of light with direction when measured on the surface of the earth.

Contrary to these ether theories, the proposed theory hypothesizes that the speed of light depends on the local spatial density, or equivalently, the local gravitational field. Since the earth's gravitational field accompanies the earth as it orbits the sun, the speed of light on the earth's surface should not be affected by the earth's velocity around the sun. However, since the hypothesized gravitational field does not rotate with the earth, the speed of light when measured on the earth's surface should vary with direction because of the earth's rotational speed. Hayden and Whitney [11] discuss experiments by Sagnac and Michelson with collaborators that indeed show the speed of light on the earth to be affected by the earth's rotational velocity, but not by its velocity around the sun. This result also agrees with Beckmann [12] who proposed that the speed of light is dependent only the gravitational field, and that the earth's gravitational field does not rotate with the earth.

9. Dynamic Electron Force and Currents

Dynamic electron force. If a hypothesized electron travels along its axis at velocity V through a flow of spatial fluid moving upward at velocity w, then a dynamic side force is generated in the plane of the electron ring. Since an electron ring rotates counter clockwise, this side force is directed to the right, in accord with the right-hand rule. If α is the angle between the ring axis and the resultant velocity, then $\tan\alpha = w/V$, as shown in Fig. 18.

This side force is analogous to the lift force acting on an uncambered airfoil placed at an angle of attack α. Using airfoil-like nomenclature, the side force is $F = C_{l\alpha}\alpha A(\rho V^2/2)$ where $C_{l\alpha}$ is a dimensionless coefficient that depends on electron ring geometry and ring circulation, A is the side-projected area of the electron ring, and ρ is density of the spatial fluid. Lumping the electron characteristics into a constant K_e, the side force is $F = K_e\alpha\rho V^2/2$.

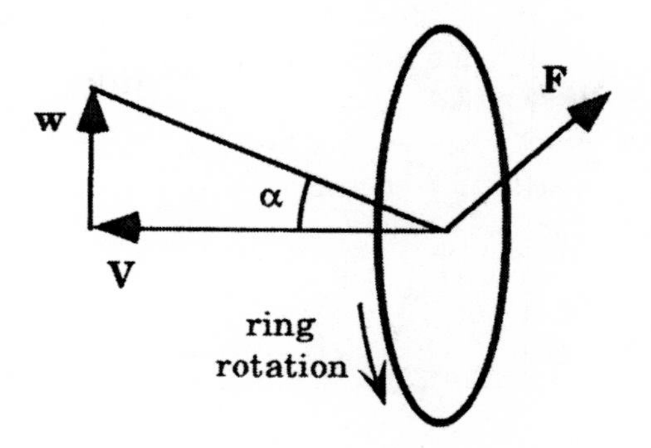

Figure 18. Dynamic electron force. The ring is shown tilting slightly to the right.

In the arbitrary case where the spatial fluid flows at angle θ relative to the electron axis, $w = W\cos\theta$. Also, if w is small relative to V, then $\alpha \approx \tan\alpha = W\cos\theta / V$. Therefore,

$$F \approx K_e \frac{W\cos\theta}{V}\frac{\rho}{2}V^2 = \frac{\rho K_e}{2} VW\cos\theta = \frac{\rho K_e}{2}\mathbf{V}\times\mathbf{W} \quad (35)$$

This force is the electromagnetic force acting on an electron traveling at V through magnetic flux density **B** as

$$\mathbf{F}_e = q_e\mathbf{V}\times\mathbf{B} \quad (36)$$

Equating (35) and (36) provides K_e expressed in terms of q_e. Note the analogies between W and B, and between K_e (which includes electron ring circulation) and electron charge q_e.

$$K_e \approx q_e(2/\rho)(B/W) \quad (37)$$

Electric currents in wires. An electric current flowing along a wire induces a clockwise magnetic field when looking in the direction of the current. It is hypothesized that each moving electron causes a "pinned" proton in the wire to orient parallel. Because of electron movement, the stationary, clockwise-rotating protons induce a clockwise flow of spatial fluid (magnetic field) around the wire, as shown in Fig. 19.

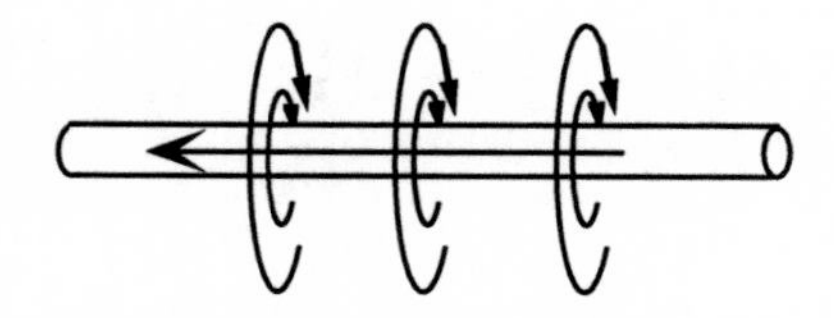

Figure 19. Flow field induced around a current-carrying wire.

Two nearby parallel wires that carry electric current in the same direction are known to attract each other, as shown in Fig. 20. The protons in each wire produce a clockwise flow of spatial fluid around the wire, inducing a fluid velocity on the other wire. This induced velocity generates a force on the counter-clockwise-rotating electrons causing the wires to attract. Similarly, any fluid flow (*i.e.*, magnetic field) perpendicular to a current-carrying wire will generate a side force on the moving electrons and wire. Electric motors are based on this principle.

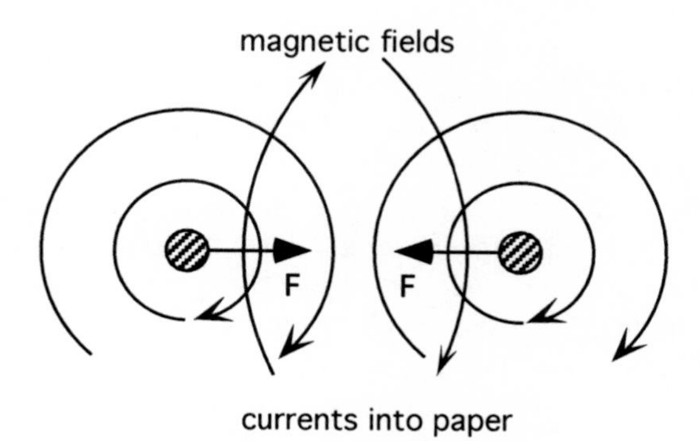

Figure 20. Force on wires, parallel currents.

Also, a current-carrying coil of wire (electromagnet) will induce a fluid flow through its center, caused by the clockwise flow of spatial fluid around each coil, which generates a return flow of fluid outside of the coil. The outer flow pattern is analogous to the magnetic field of a natural magnet. The sign convention is such that the north end of a magnet is the end from which the fluid emanates.

10. Quantum Phenomena

Quantum theory addresses the dual wave-particle nature of energy and matter, and the discreteness of matter and energy. Matter and energy are known to behave at times like waves in space, and at other times like particles. The proposed theory explains quantum phenomena because it hypothesizes that matter and energy not only behave like particles and waves, but physically exist as particles and/or waves.

An important aspect of quantum mechanics deals with the diffraction of photons. Experiments show that, if a large number of photons are directed toward a wall containing parallel sharp-edged slits, then photons that pass through the slits are deflected into diffraction patterns as if they were waves. Similarly, if a large number of photons are directed one-at-a-time toward the slotted wall, experiments show that they generate similar diffraction patterns, indicating that each photon passes through all of the slits.

This result is conceptually explained by the proposed theory. If, for example, a single photon approached one of the slits, and reached its "particle" state within one wavelength of the front side of the slit, then the photon would enter its "wave" state (while acting as a fluid source) as it continues to move toward the slit. Midway through its cycle, the photon would have fully transformed into its wave state. The photon would then begin to transform back into its particle state while acting as a fluid sink. The photon would have passed through the slit when it again reached maximum physical size. While growing after passing through the slit, the photon receives fluid that passes through all of the slits. This fluid is drawn in along different paths from the different slits, causing it to be deflected from its original path. The resulting photon path is a function of many parameters including photon wavelength, slit spacing, slit width, phase, and proximity to a slit edge when passing through.

11. Summary and Conclusions

Protons have a mass 1836.15 times that of an electron, and are hypothesized to result from the sequential amalgamation of 918 electrons and 919 positrons. Protons are found to be geometrically and dynamically similar to hypothesized positrons. A neutron is hypothesized to consist of a proton and an e-3 ring. An e-3 ring results from the amalgamation of one positron and two electrons, and behaves like an electron. Magnetic fields are motions of the spatial fluid induced by electrons and positrons. Gravitational fields result from reductions in spatial pressure caused by the reduction in spatial density surrounding all mass and energy. The dynamic force acting on electrons traveling through moving spatial fluid is analogous to their movement through a magnetic field. Also, it is shown that the proposed theory physically explains aspects of quantum mechanics. Conclusions for Part II are:

- All fields, forces, energy, and matter are hydrodynamic properties of the spatial fluid.
- Protons are formed from equal numbers of electrons and positrons, except for one extra positron.
- Electrons, protons, e-3 rings and other particles are rings that rotate at light speed and have spinning cores.
- A neutron consists of a proton and an e-3 ring that rotate in the same direction about a common axis.
- Electric and magnetic fields are movements of the spatial fluid produced by electron and proton vortices.
- Movement of an electron through a flow of spatial fluid is analogous to its movement through a magnetic field.
- Gravitational fields are gradients in spatial pressure caused by the presence of matter and energy.
- The proposed theory integrates electromagnetic fields, gravitational fields and quantum phenomena.

Acknowledgments

Comments from the reviewer and from Dr. Cynthia K. Whitney helped improve this paper, and are greatly appreciated.

References

[7] T.G. Lang, "Proposed Unified Field Theory, Part I", Galilean Electrodynamics **11**, 43-53 (2000).

[8] D.L. Bergman, "Spinning Charged Ring Model of Elementary Particles", Galilean Electrodynamics, **2**, 30-32 (1991).

[9] J.M. Blatt & V.F. Weisskopf, **Theoretical Nuclear Physics** (Dover Publications, New York, 1991; originally published by Springer Verlag in 1979).

[10] A.A. Michelson & E.M. Morley, "On the Relative Motion of the Earth and the Luminiferous Ether", Am. J. Sci. Third Series, **34**, (203) 333-345 (1887).

[11] H.C. Hayden & C.K. Whitney, "If Sagnac and Michelson-Gale, Why Not Michelson-Morley?", Galilean Electrodynamics **1**, 71-77, (1990).

[12] P. Beckmann, **Einstein Plus Two**, (Golem Press, P.O. Box 1342, Boulder, CO 80306, 1987).

APPENDIX III. PROPOSED UNIFIED FIELD THEORY - PART III; PARTICLE PROPERTIES, ATOMS AND OTHER TOPICS

Original version submitted 3 July 2000 by Thomas G. Lang. Reproduced from Galilean Electrodynamics, 14, 51-56, (2003) with permission from the publisher.

This paper is part of a series whose primary objective is to propose a new physical theory that offers a deeper understanding for the underlying mechanisms of physics. In this theory, space consists of a variable-density fluid having properties that can account for the origin, geometry, characteristics and interrelationships of energy, matter and fields. The present paper includes the theory for the formation of ring-shaped electrons, characteristics of particle rings, density of the spatial fluid, atoms, and other topics, including an overall summary of the whole theory.

Introduction

This paper is the third in a series, so numbers for Eqs., Figs., and Refs. continue from Parts I and II. In Part I [7], a photon is hypothesized to be a disturbance in space that: **1)** travels at the local speed of light, **2)** cycles smoothly between particle and wave states at a frequency proportional to its energy, and **3)** exists in its particle state as a sphere of spinning, compressed spatial fluid. An electron is hypothesized to be a structure that: **1)** forms from two photons that approach each other, hydrodynamically attract and orbit, then spread out along their optical paths to form a ring, **2)** cycles between particle and wave states, and **3)** exists in its particle state as a ring of compressed spatial fluid that rotates at the speed of light, and has a core that spins counter clockwise relative to the direction of rotation. Positrons are identical to electrons, except for opposite core spin. Electrons and positrons are hypothesized to form in equal numbers. Electron ring rotation creates the forces usually interpreted in terms of electronic charge.

Part II [13] includes the formation and characteristics of protons and neutrons, electromagnetic fields, gravitational fields, and treats topics such as electron dynamics and currents. Protons are hypothesized to form from equal numbers of electrons and positrons, plus one extra positron; they are geometrically and dynamically similar to positrons. A neutron comprises a proton and a newly-introduced e-3 ring, each rotating in the same direction about a common axis in the same plane. All charged particles are hypothesized as geometrically similar rings of compressed spatial fluid that rotate at the speed of light, and have spinning cores. All neutral particles are combinations of ring-shaped charged particles. Electric and magnetic fields are movements of the spatial fluid caused by vortices induced by electrons, protons and other charged particles. The force on an electron moving through a 'magnetic' field is caused by interaction with spatial fluid. Gravitational fields are gradients in spatial pressure caused by the presence of matter and energy. This theory unifies electromagnetic and gravitational fields, and predicts many quantum features observed in elementary particles.

The present paper develops the theory for the hypothesized creation of an electron from two photons, then analyzes the attraction force that holds electron rings together. It then deals further with protons, atoms, and other more general topics.

12. Electron Formation, Spatial Density, and Particle Densities

Initial stage of electron formation. In Part I, it was hypothesized that an electron is created from two identical, in-phase, gamma-ray photons that approach each other, attract and orbit. The centrifugal force due to orbiting is:

$$F_{cf} = m_{phe}^{*} c^2 / R_o \tag{38}$$

where m_{phe}^{*} is the mass of each formative photon at any given instant while both act as either sinks or sources, and R_o is the instantaneous orbit radius. If ρ is the density of the spatial fluid, then the attraction force between two equal three-dimensional (3-d) sinks or sources [14] is:

$$F_{ss} = \rho s^2 / 16\pi R_o^2 \tag{39}$$

The source and sink portions of a cycle are similar. Selecting the sink cycle portion, photon sink strength is:

$$s = dVol / dt = \frac{1}{\rho} d(m_{phe}^{*}) / dt \tag{40}$$

Equating (38) and (39), substituting (40), and integrating, $m_{phe}^{*} = 4\pi\rho R_o c^2 t^2$. Thus, during growth, photon mass increases with the square of time. When $t = T_{phe} / 2$, where T_{phe} is the period of the electron-forming photons, then $m_{phe}^{*} = 2m_{phe}$, where m_{phe} is average photon mass when kinetic energy equals wave energy. Substituting, $2m_{phe} = 4\pi\rho R_o c^2 T_{phe}^2 / 4$.

Substituting (4), $2m_{phe} = \pi\rho R_o c^2 h^2 / E_{phe}^2$.

Substituting $m_{phe} = E_{phe} / c^2$ and $E_e = 2E_{phe}$:

$$R_0 = E_e^3 / 4\pi\rho c^4 h^2 \tag{41}$$

The radius R_o is transitory since it is hypothesized that the orbiting photons incrementally spread out along their orbital path, a little more each cycle, until they merge, forming an electron ring. While merging, it is hypothesized that frequency and orbital radius of the formative photons smoothly increase to the final steady-state electron values.

Electron ring attraction force. We now analyze the attractive force that holds an electron ring together. Each element of an electron ring is attracted to all other elements of the ring with a hydrodynamic 'sink-sink' force during growth and a 'source-source' force when shrinking. The two phases are analytically similar, so we again address only the sink (electron growth) phase. The sink-ring analysis is complex, but the solution is known for a vortex ring. Therefore, this problem is solved by analogy between vortex rings and sink rings. From Milne-Thompson [15] and others, the mutual velocities induced by parallel line vortices and by parallel line sinks are, respectively:

$$V_\Gamma = \Gamma / 2\pi a$$

and

$$V_s = (dVol' / dt) / 2\pi a$$

where Γ is the circulation per unit length of a line vortex, $dVol'/dt$ is the sink strength per unit length of a line sink, and a is the distance between line vortices or sinks. The attraction force per unit length between parallel line vortices and sinks is:

$$F'_\Gamma = \rho\Gamma^2 / 2\pi a$$

and

$$F_s^{'} = \rho(dVol' / dt)^2 / 2\pi a$$

Notice that $dVol'/dt$ is treated like Γ in both cases. The theoretical velocity induced by a circular vortex on itself is given by Eq. (11) in Ref. 7 as $V_e = \left(\Gamma_{2dce} / 4\pi R_e\right)\left[\ln\left(8R_e / R_{ce}\right) - 1/4\right]$. This induced velocity generates an attraction force per unit length of a circular vortex, given by Eq. (15) in Ref. 7 as $F_\Gamma' = \left(\rho\Gamma_{2dce}^2 / 4\pi R_e\right)\left[\ln\left(8R_e / R_{ce}\right) - 1/4\right]$. By analogy, we replace Γ with $dVol'/dt$ to obtain the attraction force per unit length of a circular sink:

$$F_s' = \frac{\rho\left(dVol'/dt\right)^2}{4\pi R_e}\left[\ln\left(\frac{8R_e}{R_{ce}}\right) - \frac{1}{4}\right] \tag{42}$$

The centrifugal force per unit length of a rotating electron ring is:

$$F_{CF}' = \frac{m_e}{2\pi R_e}\frac{c^2}{R_e} \tag{43}$$

By definition, $dVol'/dt = (dm_e^* / dt) / 2\pi R_e\rho$. Substituting this expression into (42), and equating to (43):

$$F'_x = \frac{\rho(dm_e^* / dt)^2}{4\pi R_e(2\pi R_e\rho)^2}\left[\ln\left(\frac{8R_e}{R_{ce}}\right) - \frac{1}{4}\right] = F'_{CF} = \frac{m_e}{2\pi R_e}\frac{c^2}{R_e} \tag{44}$$

Rearranging, taking square roots, integrating and squaring:

$$m_e^* = \frac{2\pi^2\rho R_e c^2 t^2}{\ln\left(8R_e / R_{ce}\right) - 1/4} \tag{45}$$

Note that electron mass increases as t^2 during core growth. When $t = T_e / 2$, the electron reaches its maximum mass where $m^*_e = M_e$. Substituting $M_e = 2m_e = 2E_e / c^2$ and $T_e / 2 = h / 2E_e$, and rearranging:

$$R_e = \frac{4E_e^3}{\pi^2 \rho c^4 h^2}\left[\ln\left(8R_e / R_{ce}\right) - 1/4\right] \tag{46}$$

The above electron radius is greater than the transitory orbital radius of its formative photons given by (41), so the orbit radius must increase as the photons spread out along their orbit and eventually form a ring.

Spatial density. Equating (46) to Eq. (8) (from Part I; $R_e = ch/2\pi E_e$), and solving, the density of the spatial fluid is found to be:

$$\rho = \frac{8E_e^4}{\pi c^5 h^3}\left[\ln\left(8R_e / R_{ce}\right) - 1/4\right] \tag{47}$$

Maximum density of electrons. If ρ_e is the maximum density of an electron ring, and R_{ce} is electron core radius, then the maximum electron mass is:

$$M_e = \rho_e(2\pi R_e)(\pi R_{ce}^2) = 2\pi^2 \rho_e R_e^3 (R_{ce} / R_e)^2 = 2m_e \tag{48}$$

Substituting Eqs. (6) and (8) (Part I: $E_e = m_e c^2$), and rearranging:

$$\rho_e = \frac{8\pi}{c^5 h^3}\left(R_e / R_{ce}\right)^2 E_e^4 \tag{49}$$

Note that maximum electron density is proportional to E_e^4. Dividing (49) by (47), the ratio of maximum electron density to spatial fluid density is:

$$\frac{\rho_e}{\rho} = \frac{\pi^2 (R_e / R_{ce})^2}{\ln(8R_e / R_{ce}) - 1/4} \tag{50}$$

Maximum density and radius of photons that form electrons. If ρ_{phe} is the maximum density of hypothesized photons that form electrons, then their maximum mass is half of M_e, as given by (48), so:

$$M_{phe} = \frac{4\pi}{3}\rho_{phe}R_{phe}^3 = \frac{1}{2}M_e = \pi^2\rho_e R_e^3(R_{ce}/R_e)^2 \quad (51)$$

If photon density is proportional to the fourth power of photon energy, as for electrons, then

$$\rho_e/\rho_{phe} = (E_e/E_{phe})^4 = 16 \quad (52)$$

Substituting (52) into (51) and rearranging:

$$R_{phe}/R_e = (12\pi)^{1/3}\left(R_{ce}/R_e\right)^{2/3} = 3.353\left(R_{ce}/R_e\right)^{2/3} \quad (53)$$

Spin of electron cores and formative photons. Equating the maximum spin angular momentum (moment of inertia I times angular speed, ω) of two formative photons to the maximum spin angular momentum of an electron core:

$$2(0.4M_{phe}R_{phe}^2)\omega_{phe} = 0.5M_eR_{ce}^2\omega_{ce} \quad (54)$$

Substituting $M_e = 2M_{phe}$, and $R = V/\omega$ for each case, it is found that the peripheral core speed times the radius of an electron core is 0.8 that of the formative photons:

$$V_{ce}R_{ce} = 0.8V_{sphe}R_{phe} \quad (55A)$$

Alternatively, substituting (53) into (55A):

$$V_{sphe}/V_{ce} = 0.3728(R_{ce}/R_e)^{1/3} \quad (55B)$$

13. Electron and Proton Characteristics

3-d ring circulation. During each stage in the hypothesized proton formation in Part II [13], conservation of angular momentum requires that the 3-d circulation of the resulting ring remains constant, but changes sign. The 3-d circulation for incompressible fluids is defined as ring perimeter times ring velocity and effective thickness. To conserve angular momentum for a compressible spatial fluid, we must introduce a new multiplier, $(\rho_x / \rho)^n$, where subscript x represents either electron e or proton p, and n is an unknown exponent. Because the ring consists of compressed spatial fluid, it behaves as if its circulation is larger than for the case where its density is that of the surrounding fluid. Letting the effective electron thickness be $2R_{ce}k_x$ where k_x is a thickness factor, the 3-d circulation of an electron ring is:

$$\Gamma_e = -(2\pi R_e)c(k_x 2R_{ce})(\rho_e / \rho)^n \tag{56}$$

Similarly, for protons:

$$\Gamma_p = (2\pi R_p)c(k_x 2R_{cp})(\rho_p / \rho)^n \tag{57}$$

Since hypothesized electrons and protons have equal but opposite circulation, (56) is equated to -(57). By substituting (27), it is found that $(\rho_p / \rho_e)^n = (R_e / R_p)^2$. Substituting (24) and (25), it is seen that n=1/2. Therefore, (56) and (57) become:

$$\Gamma_p = -\Gamma_e = 4\pi k_x R_e^2 c(R_{ce} / R_e)\sqrt{\rho_e / \rho} \tag{58}$$

Substituting (50):

$$\Gamma_p = -\Gamma_e = 4\pi^2 k_x R_e^2 c / \sqrt{\ln(8R_e / R_{ce}) - 1/4} \tag{59}$$

The electron thickness factor, k_x, is now found by combining (59), (8), (21) and (47):

$$k_x = \sqrt{\pi\alpha} / 2 \tag{60}$$

Substituting (60) into (59):

$$\Gamma_p = -\Gamma_e = 2\pi^2 \sqrt{\pi\alpha} R_e^2 c \Big/ \sqrt{\ln\left(8R_e / R_{ce}\right) - 1/4} \tag{61}$$

Core speeds and natural speeds of protons and electrons. Since electrons and protons have identical vortex strengths, except for sign, then their core circulations must be identical, except for sign. Applying the same correction for compressibility as above, the 3-d core circulations, by definition, are core circumference times core velocity, core length and compressibility factor; *i.e.*

$$\begin{aligned}\Gamma_{ce} &= \left(2\pi R_{ce}\right)V_{ce}\left(2\pi R_e\right)\sqrt{\rho_e/\rho} = 4\pi^2 R_e V_{ce} R_{ce}\sqrt{\rho_e/\rho} \\ &= -\Gamma_{cp} = -4\pi^2 R_p V_{cp} R_{cp}\sqrt{\rho_p/\rho}\end{aligned} \tag{62}$$

Simplifying, and substituting (24), (25) and (26):

$$V_{cp}/V_{ce} = -\frac{R_{ce}}{R_{cp}}\frac{R_e}{R_p}\sqrt{\rho_e/\rho_p} = -N*N\sqrt{N^{-4}} = -1 \tag{63}$$

Therefore, proton and electron core speeds are identical, except for sign, which means that electrons, positrons and protons are geometrically and dynamically similar, except for sign.

The core spin of an electron ring vortex induces a speed along the ring axis, abbreviated 'natural speed', which was shown by Eq. (11) in Ref. 7 to be $V_e = \left(\Gamma_{2dce}/4\pi R_e\right)\left[\ln\left(8R_e/R_{ce}\right) - 1/4\right]$. The <u>2-d</u> circulation of the core of an electron ring is defined as core circumference times core speed and, corrected for compressibility as in (58) for the case of 3D circulation, is:

$$\Gamma_{2dce} = (2\pi R_{ce})V_{ce}\sqrt{\rho_e/\rho} \tag{64}$$

Substituting (64) into (11), and rearranging, the natural speed of an electron is:

$$V_e = 0.5V_{ce}\sqrt{\rho_e / \rho}\left(R_{ce} / R_e\right)\left[\ln\left(8R_e / R_{ce}\right) - 1/4\right] \quad (65)$$

Substituting (50) and rearranging:

$$V_e / V_{ce} = 0.5\pi\sqrt{\ln(8R_e / R_{ce}) - 1/4} \quad (66)$$

Equivalently, for a proton:

$$V_p / V_{cp} = 0.5\pi\sqrt{\ln(8R_p / R_{cp}) - 1/4} \quad (67)$$

Since protons and electrons are geometrically similar, the right sides of (66) and (67) are equal, giving:

$$V_e / V_{ce} = V_p / V_{cp} \quad (68)$$

Substituting (63) into (68), and ignoring the sign:

$$V_p = V_e$$

Therefore, electrons and protons have the same natural translation speed.

Approximate electron and proton core speeds. Equation (66) indicates that V_e / V_{ce} changes very little with changes in R_e / R_{ce}. For example, for values of R_e / R_{ce} of 100, 1000, and 10000, respective values of V_e / V_{ce} are 11.27, 13.14, and 14.76. Therefore, V_e / V_{ce} is around 13. Since $V_e = \alpha c = 0.00730c$, then $V_{ce} \cong 0.00056c$.

14. Atoms

Atoms consist of a nucleus of protons and neutrons surrounded at a very large distance by electrons. The number of electrons always equals the number of protons in the nucleus, and this number is called the 'atomic number' of an atom. This number determines the basic characteristics of an atom, and distinguishes the behavior of one element from another. Protons are extremely small compared with electrons. It is hypothesized that if a proton attracts an electron, then the electron is so large relative to the proton that it must accompany the proton rather than merging with it to form a more-massive particle. Therefore, the formation of atoms is the natural next step beyond the hypothesized formation of protons presented in Part II.

Whether electrons orbit the nuclei of atoms is questionable. Originally, it was believed that electrons orbit nuclei. Many now believe that static electron clouds form around the nucleus, both because there is no experimental evidence of orbiting, and because orbiting seems impossible because an orbiting electron would radiate, lose energy and spiral into the proton. Lucas [6] makes a good case for stationary electrons based on Bergman's model of an electromagnetic ring electron [4], [8]. Others, such as Whitney [16] and Klyushin [17], claim that electrons can indeed orbit a nucleus because energy losses due to radiation are compensated by other phenomena.

The theory presented here, as currently developed, assumes that electrons orbit nuclei at their natural speed, or greater. Hydrodynamically, no net energy is required for one circular vortex, such as an electron, to spiral around another circular vortex, such as a proton. Also, since the present theory assumes that electrons continuously transform between particle and wave states, and continuously change orbital paths due to the gyroscopic effect, it is expected that electrons would appear as a hazy cloud surrounding nuclei; therefore, it would seem difficult to experimentally determine whether electrons orbit nuclei or not. Whether right or wrong, it shall be assumed for now that electrons orbit nuclei.

Radius of hydrogen atoms. A hydrogen atom consists of one proton and one electron. The radius of a hydrogen atom, R_a, can now be calculated. The hydrodynamic force, F_{ee}, between an electron in orbit and a proton in the nucleus must equal the centrifugal force, F_{cf}, to maintain orbit. The hydrodynamic attraction force is identical to the so-called electrostatic force (18), so:

$$F_{ee} = q_e^2 / 4\pi\varepsilon_o R_a^2 = F_{cf} = m_e V_e^2 / R_a \tag{70}$$

Rearranging, and substituting (20):

$$R_a = \alpha hc / 2\pi m_e V_e^2 \tag{71}$$

Nils Bohr postulated that the angular momentum of an electron in orbit around a nucleus must be a multiple of the quantized angular momentum of an electron. Beckman [3] derived the same result using physical reasoning based on orbit stability. Therefore, for hydrogen, an electron completes one revolution per cycle:

$$R_a = h / 2\pi m_e V_e \tag{72}$$

Equating (71) and (72) and rearranging:

$$V_e / c = \alpha \tag{73}$$

An interesting equivalence is found between R_e / R_a and V_e / c by substituting (9), $m_e = h / 2\pi R_e c$, into (72):

$$R_e / R_a = \alpha \tag{74}$$

15. Other Topics

Neutrinos. Neutrinos cannot be detected directly, and are not well understood. When a neutron disintegrates, a proton and an electron are released, together with one or more unobservable neutrinos that are postulated to make accounts balance. If the neutron geometry hypothesized here is correct, then a neutron disintegrates into a proton and a hypothesized e-3 ring. The unstable e-3 ring then disintegrates into an electron and something that has the energy of two electrons, but has no net charge. In view of the Part I [7] Section "Photons and time", it follows that this 'something' can be conjectured as being two neutrinos that travel in opposite directions at the speed of light whereby each neutrino consists of a closely-spaced pair of out-of-phase gamma photons that spin about the travel axis in opposite directions. Linear and angular momentum are conserved, and each gamma photon has half the energy of an electron. Because of its large energy, lack of net spin, and out-of-phase component photons, the hypothesized neutrino would not significantly interact with matter, and would indeed be very difficult to detect.

Action at a distance. A twin photon experiment conducted by Nicolas Gisen at the University of Geneva, Switzerland, in September 1997 [18], provided a spectacular demonstration of the long-range connection between quantum events predicted by quantum theory. Others have since verified this result which is said to be explained by quantum theory, in that 'entangled' particles continue to communicate with each other almost instantaneously even when very far apart. Splitting a photon into two photons generates entangled photons. If one such photon is forced to undergo a change, such as in polarization, then its entangled twin undergoes a complementary change within a time period that is far less than predicted by the speed of light.

The theory proposed here provides a possible explanation. If photons cyclically expand out toward infinity and back within one photon period, as described in the Section "Photons" in Part II [7], then near-instantaneous action at a distance is explainable. Entangled photons would have identical energies, spins and phases; therefore, each photon may have enough characteristics in common to permit it to respond to a change in the other photon.

16. Equations of Physics and Hydrodynamics

The similarities between the equations of physics and hydrodynamics have been pointed out by many physicists, including Feynman [19]. As hypothesized herein, these similarities are not coincidental.

In electrostatics and magnetostatics, for example, the equations $\nabla \cdot \mathbf{B} = 0$ and $\nabla \times \mathbf{B} = \mathbf{J} / \varepsilon_o c^2$ are the same as in the proposed theory where $\nabla \cdot \mathbf{V} = 0$ and $\nabla \times \mathbf{V} = \Omega$ where **V** is the fluid velocity, and **Ω** is fluid vorticity. Fluid velocity, **V**, in the proposed theory is analogous to magnetic flux, **B**, in physics. Vorticity in the proposed theory is analogous to '**J**' in physics. The hypothesized hydrodynamic 3D circulation, Γ_e, of an electron is analogous to electron charge. The hypothesized hydrodynamic force, $F_{ee} = \rho \Gamma_e^2 / 4\pi r^2$, is analogous to electrostatic force, $F_{ee} = q_e^2 / 4\pi\varepsilon_o r^2$. Likewise, the hypothesized hydrodynamic force on an electron moving through spatial fluid is analogous to an electron moving through a magnetic field. The hypothesized hydrodynamic buoyant force between two masses, $F_G = d(\Delta P_1) / dr(m_2 / \rho)$ is the gravitational force $F_g = Gm_1 m_2 / r^2$ between two masses, where $d\Delta P_1 / dr = \rho G m_1 / r^2$.

17. Summary of Parts I, II and III

What is proposed here is a new view of the Universe in which all matter moves at the speed of light, nothing is at rest, and all matter and energy cyclically disappear and reappear. Nothing is solid. All matter is a form of energy.

On the other hand, in every-day life, matter is perceived as solid, does not disappear and reappear, can remain at rest, and is distinct from energy. However, it has long been known that atoms comprise a nucleus surrounded at a very large distance by electrons. The volume of the nucleus and electrons is so small that an atom is essentially empty. Thus, perception can be quite different from reality.

Interestingly, physicists do not know what photons look like and how they propagate through space; what an electron looks like; what protons, neutrons, and neutrinos look like; how energy and mass are physically related; what an electric charge is; how gravitational and electromagnetic fields are related; how quantum phenomena can be physically explained; and what physically provides the forces in atomic nuclei.

The objective of the proposed theory is to answer these kinds of physical questions with the help of classical hydrodynamic and physics theory. Here, mass, energy and space are hypothesized to be interchangeable, and their sum is conserved. Also, angular and linear momentums are conserved.

Space is envisioned to consist of a variable-density fluid having some special properties. Photons are hypothesized to be pure energy traveling at the local speed of light, smoothly cycling between particle and wave states at a frequency proportional to energy, with the particle state consisting of a sphere of spinning, compressed spatial fluid. Electrons are hypothesized to result from two orbiting photons, to transform smoothly between particle and wave states at a fixed frequency, and, in their particle state, to consist of a ring of compressed spatial fluid that rotates at the speed of light with a core that spins counter clockwise. Positrons are identical to electrons, except for opposite core spin.

Electron dynamics, including natural electron speed, is developed. The circulation of hypothesized electron rings is shown to be analogous to electronic charge. Protons are hypothesized to be formed from 918 electrons and 919 positrons, and are geometrically and dynamically similar to positrons. A neutron is hypothesized to consist of a proton and a newly introduced e-3 ring. The angular momentum, magnetic moment and spin of the proposed models for electrons, positrons, protons and neutrons are in agreement with experiment. All charged particles are hypothesized to be rings that rotate at the speed of light, and to be geometrically and dynamically similar. All neutral particles are hypothesized to be combinations of charged particles. This theory derives Einstein's famous equation, $E = mc^2$, using simple equations. Magnetic fields and gravitational fields are hypothesized to be motions of the spatial fluid induced by matter and energy. Relationships of proposed photons, particles, and fields to quantum phenomena have been described and explained. Equations are developed relating the proposed unified field theory with experimental phenomena. All relationships are functions of seven basic parameters, which can probably be reduced after further analysis.

18. Overall Conclusions

- All fields, forces, energy, and matter are hydrodynamic properties of the spatial fluid.
- Mass, energy and space can transform back and forth into each other.
- Photons are pulsating spheres of spinning, compressed spatial fluid that travel at the local speed of light.
- Electrons and positrons are formed from pairs of gamma-ray photons that orbit and merge into rings.
- Electrons are pulsating rings with spinning cores of compressed spatial fluid that rotate at light speed.
- Electrons and positrons are identical except for core spin direction, and are a form of energy.
- Protons are geometrically similar to positrons but are 1836 times more massive and 1836 times smaller.
- Protons are formed from equal numbers of electrons and positrons, plus one extra positron.
- An e-3 ring is similar to an electron, but has three times more mass, and is three times smaller.

- A neutron is a proton coupled with an e-3 ring that rotate at light speed about a common axis.
- All charged particles are geometrically and dynamically similar rings that rotate at the speed of light.
- The maximum density of charged particles is proportional to the fourth power of their energy.
- Charged particles typically repel, but can attract depending upon their orientation.
- All neutral particles are physical combinations of charged particles.
- Electric and magnetic fields are movements of the spatial fluid generated by charged particles.
- Gravitational fields are gradients in spatial pressure generated by the presence of matter and energy.
- Spatial pressure reduces inversely with distance from any mass or photon.
- The natural translation speed of electrons results from the spin of their cores.
- Action at a distance results from photon and particle transformations between particle and wave states.
- The red shift is conjectured to be at least partly caused by photon energy loss with distance.
- Everything moves at the speed of light, and cyclically disappears and reappears; nothing is solid.
- This theory unifies electromagnetic and gravitational fields, and physically explains quantum features.

Acknowledgements

The author thanks Dr. Cynthia Whitney, Editor of Galilean Electrodynamics (GED), for her many insightful suggestions, and professional expertise in editing the three parts of this paper, and revising the drawings. The author also thanks the anonymous GED reviewer for a careful review, finding an error in my original derivation of electron ring attraction force, for suggesting many changes that helped to augment my limited knowledge of physics, being a Ph. D. hydrodynamicist, and for reviewing all three parts of this paper even though he prefers an electromagnetic approach.

References

(Note: Refs. [3] and [6] are repeated from Part I; Refs. [7] and [8] are repeated from Part II.)

[3] P. Beckmann, **Einstein Plus Two**, pp.124-127, (Golem Press, P.O. Box 1342, Boulder, CO 80306, 1987).

[6] J. Lucas, "A Physical Model for Atoms and Nuclei", Galilean Electrodynamics, 7, 3-12 (1996).

[7] T.G. Lang, "Proposed Unified Field Theory, Part I", Galilean Electrodynamics 11, 43-48, (2000).

[8] D.L. Bergman, "Spinning Charged Ring Model of Elementary Particles", Galilean Electrodynamics **2**, 30-32 (1991).

[13] T.G. Lang, "Proposed Unified Field Theory, Part II", Galilean Electrodynamics **12**, 103-107 (2001).

[14] L.M. Milne-Thompson, **Theoretical Hydrodynamics**, 4th edition, pp. 454-471 (Macmillan, New York, 1960).

[15] Ibid., p. 180, 200.

[16] C.K. Whitney, "A New Perspective on the Hydrogen Atom", Physics Essays **1** (2) 53-55, (1988).

[17] J.G. Klyushin, "A Field Generalization for the Lorentz Force Formula", Galilean Electrodynamics **11**, 43-48, (2000).

[18] A. Watson, "Teleportation Beams Up a Photon's State", Science 278 1881-1882 (1997). (Abstract of article in Nature, 12 December 1997 by Anton Zeilinger.)

[19] R.P. Feynman, R.B. Leighton, & M. Sands, **The Feynman Lectures on Physics, Vol. II,** p. 40-5 (Addison-Wesley, Reading, MA, 1963).

APPENDIX IV. PROPOSED UNIFIED FIELD THEORY - PART IV; COMMENTARY

Original version submitted 7 August 2004 by Thomas G. Lang. Reproduced from Galilean Electrodynamics with permission from the publisher, as printed in the section "From the Editor: A letter from our files", 20, (2009).

> The objective of the proposed unified field theory is to offer a physical explanation for the underlying mechanisms of physics that leads to a deeper understanding. The theory is based on hydrodynamics. This Part IV summarizes and updates Parts I-III [1-3], which contain the illustrations, equations, and derivations.

Pending Questions

Many questions remain unanswered in physics, such as: How can $E = mc^2$ be derived physically? How can energy and mass physically transform into each other? What is electricity? What do photons, electrons, protons and neutrons look like? What physically causes electric, magnetic and gravitational fields and forces? How can electric, magnetic and gravitational fields be unified? How can photons and matter behave like both particles and waves? What is the strong force in a nucleus? How fast does gravity act? How can electrons and positrons annihilate into photons? Where are all of the positrons? What are neutrinos? How can quantum phenomena be physically described? What permits action at a distance? What permits superconductivity? Is the universe expanding? This theory addresses these questions, and others.

Space. In the proposed theory, space consists of a variable density fluid. This spatial fluid has special properties that account for the origin, geometry, characteristics and interrelationships of energy, matter and fields. The existence of a spatial fluid solves the paradox of an otherwise empty space that contains electromagnetic and gravitational fields, and quantum phenomena.

Photons. We begin with photons, which are considered to be pure energy, and are proposed here to be the building blocks for everything else. Photons are hypothesized to be disturbances in space that: **1)** travel through the spatial fluid at the local speed of light, **2)** pulse smoothly and periodically between wave and 'particle' states at a frequency proportional to their energy, **3)** exist as spheres of compressed spatial fluid in their particle states that grow from zero to a maximum and back again during each cycle, (4) act like hydrodynamic sinks that draw in spatial fluid when growing, (5) act like hydrodynamic sources that return the spatial fluid when shrinking, (6) spin about their travel axis one way or the other in equal numbers, and (7) are accompanied by local periodic reductions in spatial density and pressure that are proportional to their energy, and inversely proportional with distance.

Energy. The photon properties postulated above suggest that energy is a special form of spatial fluid.

Electrons and Positrons. Electrons and positrons are known to be the lightest stable form of matter. These particles are hypothesized to result from two gamma photons, each having half the energy of an electron. If two such photons have the same frequency, phase and spin direction, and approach each other closely enough, then they will hydrodynamically attract and orbit because of the known hydrodynamic attraction force between two sources or two sinks. Once in orbit, it is hypothesized that the photons spread out along their common path until they form a ring. Assuming that momentum and energy are conserved during ring formation, each such ring will rotate at the speed of light, and its core will have the same spin direction and angular momentum as the formative photons. Like the formative photons, this core will pulse smoothly from zero to a maximum thickness and back again at a frequency that is proportional to ring energy. Rings whose cores spin counter-clockwise in the direction of ring rotation are defined as electrons; while those that spin clockwise are called positrons. Assuming that photon spin is equally distributed each way, equal numbers of electrons and positrons are formed. Although these rings rotate at the speed of light, they can remain at rest, unlike photons, and are therefore considered particles of matter, and have a rest mass.

Matter. The above discussion suggests that matter is a special form of energy.

Density and Pressure Fields. Like photons, the postulated electrons and positrons are accompanied by reductions in spatial density and pressure that vary inversely with distance. Therefore, both energy and matter are accompanied by fields of reduced spatial density and pressure that are proportional to their energy or mass, and inversely proportional to distance from each element of energy or mass. These density and pressure fields cycle at the same frequency as the corresponding photon or particle, and extend outward toward infinity and back like sources and sinks. These fields must therefore change much faster than the speed of light.

Equivalence of Matter and Energy. On the average during any cycle, half of the energy, E, of a ring particle is wave energy, and half is kinetic energy. The average ring kinetic energy is $\frac{1}{2} mv^2$, where m is the average particle mass during one cycle, and v is the ring velocity. Since ring velocity is the speed of light, c, the average ring kinetic energy is $\frac{1}{2} mc^2$. Since the total ring energy is twice the average kinetic energy, then $E=mc^2$, which is the famous Einstein relationship. Note how easily this relationship is derived using the proposed physical model.

Particle-Wave Paradox. The proposed theory suggests that photons and matter physically transform between particle and wave states. The diffraction grating paradox is solved because a hypothesized electron ring or photon can follow a particle-like trajectory on each side of a diffraction grating, but within one wavelength of the grating, their periodic wave-like properties permits them to simultaneously pass through all of the slits [2].

Annihilation. It is known that electrons and positrons annihilate into gamma photons. Mass is converted into energy. Annihilation is discussed in [1], showing physically how and why gamma photons are released, and why annihilation is essentially the reverse of particle formation.

Red Shift. We have presented different ways in which spatial fluid, energy and mass can transform back and forth into each other. Missing, is the conversion of energy back into space. It is conjectured that a tiny amount of energy is transferred back into spatial fluid in the form of increased spatial density and pressure as a photon travels through space, thereby reducing its frequency. If this conjecture is so, then either a part, or possibly all, of the well-known red shift in light received from remote sources is caused by a reduction in photon energy with distance. Consequently, the universe may either be not expanding, or expanding slower than currently believed.

Missing Positrons. If electrons and positrons are generated in equal numbers, as hypothesized, then where are the missing positrons? The answer below is that each proton contains a missing positron. Since the number of electrons and protons in the universe are equal, the missing-positron paradox is solved.

Protons. It is known that protons are 1,836.15 times more massive than positrons or electrons. It is hypothesized that 919 positrons combine with 918 electrons to form a proton. The difference from the combined mass is binding energy. Proton growth is hypothesized to occur in stages, as described later.

Electric Force. Hydrodynamic theory shows that a force is generated on a vortex that is placed in a flow of fluid. This force is perpendicular to both the fluid flow and axis of the vortex, and is directed from the vortex center toward the side where the vortex flow parallels the fluid flow. In the present theory, any rotating particle ring acts like a vortex, so any relative motion of the spatial fluid will cause an electric-like force.

Ring-Ring Forces. Hydrodynamic theory shows that counter-rotating rings lying in a plane will attract, while rings that rotate in the same direction will repel. Each rotating ring induces a fluid flow on the other ring such that an "electric force" is generated, as discussed above. It is further shown in [1] that two isolated rings will dynamically interact until they "point' in the same direction, and will therefore lie in the same plane. A ring is herein defined to "point' in the direction of flow induced through its center by core rotation.

e-3 Rings. The first stage in proton growth is hypothesized to begin when a positron ring attracts two electron rings, one on each side, each lying in the same plane. As the electron rings approach and contact the positron ring, their rings will mesh, somewhat like gears, preventing annihilation. Once meshed, the electrons are envisioned to hinge outward in opposite directions until pivoting through 180 degrees, at which point each electron ring fully meshes with the positron ring, forming a new ring-shaped particle. Assuming conservation of mass, momentum and energy, this new ring will have three times the mass of an electron, one-third its diameter, the same core spin, and the same ring rotation speed [2]. This new particle is called an e-3 ring because it dynamically acts like an electron, but has three times its mass.

Proton Formation. Independently, an electron is hypothesized to attract two positrons to form a p-3 ring that acts like a positron, but has three times its mass. The second stage in proton formation consists of similarly forming e-9 and p-9 rings from the e-3 and p-3 rings. These stages continue until a p-1837 ring is formed, which is considered to be a proton. A proton dynamically acts like a positron, but is about 1837 times more massive and smaller. This process explains why so many particles can result from the annihilation of a proton since any particle, or combination of particles, formed during the formation of a proton can result from its annihilation. Also, neutral particles can result from combinations of charged particles.

Particle Density. Ref. 3 shows that the density of the hypothesized photons and ring particles, when at their maximum particle size during any cycle, is proportional to the fourth power of their energy. Therefore, gamma photons, and electron rings, because of their high energy, are predicted to have very large maximum densities. Similarly, because of their even greater energy, protons and neutrons will have much larger maximum densities. This result helps to physically explain the large densities of neutron stars and black holes.

Atoms. Atoms have a nucleus consisting of protons and neutrons that is surrounded at a large distance by electrons. There are an equal number of protons and electrons in an atom, and this number determines the characteristics of each element. Whether electrons orbit the nuclei of atoms is questionable. Many physicists now believe that static electron clouds form around the nucleus because there is no experimental evidence of orbiting, and because orbiting is considered impossible because an orbiting electron would radiate, lose energy and spiral into the nucleus. However, other physicists have shown that electrons can orbit a nucleus without losing energy. The theory presented here, as currently developed, assumes that an electron orbits a nucleus at its natural speed, or greater, depending on whether energy is added to the electron.

Natural Ring Speed. Because of their core spin, the proposed rings have a natural speed, much like the natural speed of a smoke ring. This speed is hydrodynamically induced by core spin. As shown in [3], this natural speed is $c/\alpha = c/137.037$, where α is the fine-structure constant.

Neutrons. Neutrons have no charge, and have approximately the mass of one proton, plus three electrons. It is hypothesized that a neutron consists of a proton ring and an e-3 ring that rotate in the same direction about a common axis in the same plane. This geometry is hydrodynamically marginally stable, which explains why an isolated neutron can remain intact for as long as 15 minutes on average. The binding energy of a neutron must therefore be very small. The measured electric field of a neutron tends to support this model.

Neutral Particles. A significant result of the proposed theory is that all neutral particles are combinations of charged particles.

Proton-Neutron Pairs. In view of the above-hypothesized neutron, a proton-neutron pair should consist of an e-3 ring that lies between two proton rings, each rotating in the same direction about a common axis. The cores of the three rings will lie close to each other because the e-3 core is hydrodynamically attracted to the two proton cores since it rotates in the opposite direction. This attraction, as shown in [2], causes the diameter of the e-3 ring to greatly reduce, approaching the diameter of the proton rings. The proton rings expand very slightly under the same attractive force because they are much less elastic. A proton-neutron pair should be very stable because the proton cores repel, while being strongly attracted to the central e-3 core.

Ring elasticity. Each hypothesized ring consists of compressed fluid, and is therefore elastic. The small, very-massive proton rings are held together by much stronger hydrodynamic forces than the large, low- mass e-3 rings. Ref. 3 shows that ring elasticity is proportional to the square root of ring mass. Therefore, the above-mentioned attractive force will shrink an e-3 ring far more than the proton rings expand, resulting in a geometry where the cores of the three rings will lie very close together, and within a proton radius.

Strong Force. These core-core forces are orders of magnitude larger than normal electric forces because of the short core-core distances compared with typical distances between electric charges. It is hypothesized that this core-core force is the "strong force" in theoretical physics that binds nuclear particles together. This new model of the nucleus provides a physical reason why protons and neutrons are bound together so strongly, while isolated neutrons are so lightly bound that they are marginally stable.

Magnetic moment, moment of inertia and spin. The calculated magnetic moments, angular moments of inertia and spins of the proposed ring particles are shown in [1] and [2] to agree with measured values.

Magnetic Fields. Magnetic fields result from the core spins of electron and proton rings. For example, if equal numbers of electrons and protons align to form a bar magnet, then the flow induced by ring rotation cancels, leaving only the flow induced by the spins of their cores. Since the cores of the aligned rings spin in the same direction, a strong flow of spatial fluid is induced through their centers, with the return flow lying outside of the aligned rings. This flow pattern can be seen by sprinkling iron filings on paper placed above a bar magnet. The flow directions are such that the central flow leaves from the north end of the magnet.

Gravitational Fields. A gravitational field in this unified field theory is simply the pressure field that results from the presence of matter and energy. As discussed earlier, spatial pressure reduces proportional to mass, and varies inversely with distance. Photons act similarly, except an equivalent mass is used where $m = E/c^2$. Gravitational force is considered a "buoyant-like" pushing force, like buoyancy in water, calculated as the product of the local pressure gradient and "equivalent volume" of the associated mass,. The "equivalent volume" is defined as mass divided by the local spatial density. For example, the gravitational force between two masses is shown in [2] as being proportional to the product of the masses, and inversely proportional to the square of distance between them. A gravitational force acts almost instantaneously because the hypothesized reduction in spatial pressure takes place within one photon or particle cycle, as discussed earlier.

2-d and 3-d Electric Fields. As proposed herein, fluid fields and hydrodynamic forces are analogous to electric fields and electric forces. For example, hydrodynamic theory shows that two rotating rings offset sideways in a fluid will attract if the rings rotate in opposite directions, and will repel if they rotate in the same direction. Therefore, a hypothesized electron ring and a proton ring will attract, while two electron rings or two proton rings will repel. The proposed theory works well if the rings lie in the same plane, or if they lie within a few diameters of each other in three-dimensional space, such as in a nucleus. The proposed theory can be made to work equally well in all other three-dimensional cases by introducing a fourth-dimensional property (which could be called a "quantum effect", or an "action at a distance effect") wherein any pair of rings act as if they lie in the same plane.

Electron Orbits. The proposed electron ring model provides a physical explanation for electron clusters and electron coupling. For example, two of the hypothesized electron rings can orbit stably about a central axis since they will point in opposite directions and therefore attract, as discussed in [2] and [3]. Similarly, many electrons could orbit about a single axis. These electrons could also orbit an atomic nucleus, in which case the added attraction force from protons in the nucleus would reduce the orbit diameter. Likewise, two sets of orbiting electrons could stably orbit off-center around a nucleus because their attraction to protons in the nucleus would balance the opposing force between the orbiting sets. Furthermore, additional sets of electrons could orbit a nucleus at greater distances in the heavier atoms.

Alternative Electron Pairing. A second type of electron pairing appears possible wherein two electron rings lie in closely-spaced parallel planes, point in opposite directions, and orbit rapidly about an axis lying in a central plane. Their opposite-rotating cores would attract, balancing the centrifugal force.

A third type of electron pairing may be possible, similar to an unusual hydrodynamic phenomenon wherein a pair of ring vortices can travel in tandem, and continually trade places. The trailing vortex is hydrodynamically attracted to the lead vortex, and while catching up, its diameter reduces relative to the lead vortex diameter, permitting it to pass through the inside of the lead vortex ring, where it now becomes the lead vortex ring. This dynamic behavior can repeat, over and over again. Since electron rings are ring vortices, they might behave similarly, unless their core pulsations or ring rotations interfere with this behavior.

Superconductivity. Pairs of electrons, called Cooper pairs, were found to be intimately involved in superconductivity. It is conjectured that one of the above electron-pair models could be a Cooper pair. The first two models mentioned above are the most likely because their rings rotate in opposite directions, minimizing the ring rotation effect, thereby permitting the electron pair to travel with minimum resistance.

Neutrinos. Neutrinos cannot be directly detected, and are not well understood. It is known that a neutron disintegrates into a proton and an electron, together with one or more unobservable neutrinos that are postulated to make accounts balance. If our hypothesized neutron geometry is correct, then a neutron will disintegrate into a proton and an e-3 ring. The e-3 ring then disintegrates into an electron and a positron/electron pair that could then annihilate into four gamma photons. It is hypothesized that four gamma photons could form two neutrinos that travel in opposite directions at the speed of light. If so, then a neutrino would consist of a tandem pair of gamma photons that spin in opposite directions about a common axis. The hypothesized neutrinos satisfy conservation of linear and angular momentum, and would be very difficult to detect because their lack of net spin would minimize their interaction with matter.

Time. The time taken for a photon to travel a given distance is that distance divided by the speed of light. Consequently, if the speed of light is zero, then an infinite time is needed for light to travel any given distance; no action would take place, and there would be no meaning to time. On the other hand, if the speed of light was infinite, then all action would occur simultaneously; and time would again be meaningless. Therefore, the existence of a finite speed of light permits sequencing of events that leads to the idea of time.

Action at a Distance. A twin photon experiment conducted by Nicolas Gisen at the University of Geneva, Switzerland, in September 1997, provided a spectacular demonstration of the long-range connection between quantum events involving entangled photons that occurs far faster than the speed of light, and is predicted by quantum theory. Others have since verified this result. The theory proposed here provides a possible physical explanation. If photons, in their wave state, pulse outward toward infinity and back within one photon period, as hypothesized, then near-instantaneous action at a distance is possible. Entangled photons would have identical energies, spins and phases, and might have enough characteristics in common to permit one photon to respond to a change in the other within one photon period, no matter how far apart the photons might be.

Quantum phenomena. Quantum theory addresses the dual wave-particle nature of energy and matter, and the discreteness of matter and energy. Therefore, many aspects of the proposed theory fall under this definition, including the cyclical appearance and disappearance of photons and ring particles, the reduction in spatial density and pressure that accompanies all energy and matter, the resulting gravitational field, and action at a distance. Consequently, the proposed theory could be said to unify quantum phenomena with electromagnetic fields and gravitational fields.

Hydrodynamic and Electromagnetic Fields. Similarities in equations between these fields have often been noted in the literature. These similarities may not be coincidental, as discussed in [3].

Questions Revisited. Interestingly, physicists have developed equations that predict most of the known physical processes. However, physicists do not know what photons physically look like or how they propagate through space; what electrons, protons, neutrons, or neutrinos physically look like; how energy and mass are physically related; how to explain the wave–particle paradox, what an electric charge physically is; how gravitational and electromagnetic fields are physically related, how these fields can exist in empty space; how to physically explain quantum phenomena; and what causes the strong force in nuclei.

Answers. The proposed theory offers answers to these questions. Here, mass, energy and space are hypothesized to be interchangeable, and their properties conserved. Nothing is created or destroyed. A new view of the universe is proposed here where all matter moves at the speed of light, nothing is at rest, and all matter and energy cyclically disappears and reappears. Nothing is solid. All matter is a special form of energy, and energy is a special form of spatial fluid.

On the other hand, in every-day life, matter is perceived as solid, does not disappear and reappear, can remain at rest, and is distinct from energy. However, it has long been known that atoms comprise a nucleus surrounded at a very large distance by electrons. The volume of the nucleus and electrons is so small that an atom is essentially empty. Thus, perception can be very different from reality.

Summary

- All fields, forces, energy, and matter are hydrodynamic properties of the spatial fluid.
- Mass, energy and spatial fluid can transform into each other. Nothing is created or destroyed.
- Photons are pulsating spheres of spinning, compressed spatial fluid that travel at the local speed of light.
- Electrons and positrons are formed from pairs of gamma-ray photons that orbit and merge into rings.
- Electrons are pulsating rings with spinning cores of compressed spatial fluid that rotate at light speed.

- All ring particles have a natural speed that results from the spin of their cores.
- Electrons and positrons are identical except for core spin direction.
- Protons are geometrically similar to positrons, but are 1836 times more massive and 1836 times smaller.
- Protons are formed from equal numbers of electrons and positrons, plus one extra positron.
- All charged particles are geometrically similar rings with spinning cores that rotate at the speed of light.
- An e-3 ring acts like an electron, but has three times its mass, and is three times smaller.
- All neutral particles are combinations of charged particles.
- A neutron consists of a proton and an e-3 ring that rotate in the same plane about a common axis.
- The maximum cyclic density of a charged particle is proportional to the fourth power of its energy.
- Similarly-charged particles typically repel, but they can attract in special situations.
- Electrons can orbit either independently, around a nucleus, or off-center from a nucleus as two orbits.
- Electric and magnetic fields are movements of the spatial fluid caused by charged particles.
- A reduction in spatial pressure accompanies all mass and energy, and falls off inversely with distance.
- The spatial pressure field is the gravitational field. A pressure gradient causes a gravitational force.
- Action at a distance results from the cyclical pulsation of particles and photons toward infinity and back.
- The red shift is at least partially caused by photon energy loss with distance.
- Everything moves at the speed of light, and cyclically disappears and reappears; nothing is solid.
- The proposed theory unifies electromagnetic fields, gravitational fields and quantum phenomena.

References

[1] T.G. Lang, "Proposed Unified Field Theory, Part I", Galilean Electrodynamics 11, 43-48, (2000).
[2] T.G. Lang, "Proposed Unified Field Theory, Part II", Galilean Electrodynamics, 12, 103-107 (2001).
[3] T.G., Lang, "Proposed Unified Field Theory, Part III", Galilean Electrodynamics, 14, 51-56, (2003).

GLOSSARY, AND SECTION INDEX

(The first number in each parenthesis relates to a section in the main part of the book, while numbers that start with "A" relate to a section in the first three appendices.)

Atoms (7, A14): Atoms in this new theory remain much like those in modern physics. However, here the nucleus consists of *proton rings* and neutrons, which is *orbited by electron rings*. Similar to modern physics, the number of electron rings (electrons) always equals the number of proton rings (protons), and each such number determines the characteristics of each of the 92 elements found in nature, starting with hydrogen. Elements beyond these 92 are manufactured, and are radioactive.

Big Bang (11.2, A4): The hypothesized explosion that started our expanding universe 13.7 billion years ago, according to the prevailing theory in modern physics.

Black Holes (12.2): The remnants of stars that ran out of hydrogen fuel, collapsed, and are so massive that their gravity prevents light from escaping.

Core-Core Forces (5.4, A7): Forces between the spinning cores of adjacent fluid rings that are thousands of times stronger than electric forces. Core-core forces are hypothesized in this new theory to be the nuclear strong force in modern physics.

E-3 Ring (5.3, A7): A hypothesized fluid ring that acts much like an electron ring, but has three times its mass, and is 1/3 its size.

Electron Pairs (10.1, A4): The ability of two hypothesized electron rings to fluid-dynamically pair by orbiting around a central axis.

Electron Rings (5.1, A3): A hypothesized fluid ring that rotates at the speed of light, and has a pulsing core that spins counter-clockwise in the direction of rotation. A positron ring is identical, except for a clockwise core spin.

Ether Theory (3.2, A8): A theory based on an ether medium that filled all space, through which the earth and everything else moves. This theory was widely accepted in the 1800's, but was abandoned following an experiment conducted by Michelson and Morley in 1887 that showed that the earth did not move through any such ether.

Fluid Rings (5.2, A7): A name given to all hypothesized rings of spatial fluid that rotate at the speed of light, have pulsing cores that spin one way or the other, and are geometrically and dynamically similar. Electron and proton rings are different types of fluid rings. (Note that all charged particles in modern physics are considered to be fluid rings in this theory.)

Gravity (9, A8): Gravity is hypothesized to be a pressure force that is caused by pressure gradients in the spatial fluid that result from a basic pulsing property that is shared by all photons and particles of matter.

Magnetic Fields (6.6, A8): Hypothesized fluid-dynamic fields that result from aligning electron rings and proton rings. A bar magnet is an example of such an alignment.

Modified Big Bang (11.2): A type of Big Bang universe that is modified to include the effects of a hypothesized new photon energy loss with travel distance.

Mysteries in Modern Physics (2, A1): All phenomena in modern physics that cannot be physically understood.

Neutrinos (5.5, A15): Neutrinos travel at the speed of light, cannot be directly detected, and are not well understood. Here, neutrinos are hypothesized to consist of two closely spaced, opposite-spinning gamma photons that travel in tandem at the speed of light.

Neutrons (5.4, A7): Neutrons are here hypothesized to consist of an inner proton ring and an outer e-3 ring that lie in the same plane, point in the same direction, and rotate in the same direction about the same axis. (In this new theory, all neutral particles consist of combinations of fluid rings, and here, various kinds of fluid rings are equivalent to charged particles in modern physics.)

Neutron Stars (12.2): Remnants of stars that ran out of hydrogen fuel, exploded as a supernova while losing about half of their mass, and consist primarily of neutrons.

New Stars and Galaxies (12.4): Observations indicate that new stars and galaxies are formed continuously, primarily from aggregations of space dust and hydrogen.

Nuclear Strong Force (7.2, A7): Hypothesized here to be the core-core force that acts between nearby spinning cores of fluid rings, and is thousands of times stronger than electric forces.

Nuclear Weak Force (7.4): The force that causes neutrons to transform into protons, and vice-versa. In this theory, a neutron consists of a proton ring that lies inside, and concentric to, an e-3 ring. This geometry is marginally stable because only a small side force is needed to move the e-3 ring far enough sideways to break up a free neutron. This side force is considered here to be the weak nuclear force because this force is involved in either forming or breaking up a neutron.

Photons (4, A2): Photons are hypothesized to travel at a local speed of light, *have a moving mass*, pulse, and spin one way or the other about their travel axes in equal numbers throughout the universe. The only difference between photons is their pulsing frequency, which is proportional to their energy.

Photon Creation (4.3, A2): Photons are hypothesized to form whenever energy is transferred into the spatial fluid. Photons form in multiples of four in order to satisfy conservation laws.

Photon Energy and Frequency (4.6, A2): Photon energy is known to be exactly proportional to photon frequency. Here, photon frequency is its pulsing frequency. Also, photon energy is hypothesized to result half from forward-moving kinetic energy, and half from the combined kinetic energy of the photon spinning and pulsing behaviors. In other words, *all photon energy is kinetic*, which is defined as being *pure energy*.

Proton Rings (5.3, A7): Hypothesized fluid rings that here result from the combination of 918 positron rings and 917 electron rings. A proton ring is geometrically similar to a positron ring, rotates at the speed of light, has the same core spin as a positron ring, but has 1836.25 times its mass, and is that many times smaller. (The small difference in mass from the total of 1837 rings is attributed to binding energy.)

Proton-Neutron Pairs (7.3, A7): Such pairs are hypothesized to consist of a central e-3 ring that is sandwiched between two smaller proton rings; all three rings rotate in the same direction around a common axis. The cores of these three rings lie very close to each other, and are strongly bound by *core-core forces*, which are considered here to be *the strong nuclear force*.

Pulsing (4.4, A2): In this theory, both photons and particles of matter are hypothesized to cyclically pulse outward and back at high frequencies. It is this pulsing property that physically provides photons and matter with their quantum properties, and permits quantum behaviors to be physically understood.

Red Shift (1.1, A4): A downward shift in the frequency

of light toward the red end of the light spectrum. A Red Shift is observed in the light that comes from stars and galaxies. The amount of Red Shift increases with travel distance. The Red Shift can result from either a Doppler Shift (as believed in modern physics) caused by recession speed, or a loss in photon energy with travel distance.

Quantum Phenomena (8, A10): Quantum phenomena here result from the hypothesized pulsing behavior of photons and matter. This newly hypothesized pulsing behavior is especially important because it provides a physical understanding of all quantum phenomena.

Ring-Ring Forces (5.2): Hypothesized fluid-dynamic forces between rotating fluid rings. These ring-ring repulsion and attraction forces depend on the direction of core spin. These two kinds of spin are closely analogous to positive and negative electric forces in modern physics.

Sources and Sinks (5.1, A3): These fluid dynamic terms are nicely described by their names. A source is a fictional point in a fluid from which fluid flows outward in all directions. A sink is a fictional point in a fluid into which fluid flows from all directions. Fluid dynamic theory shows that either two sources or two sinks attract, while a source and a sink repel. When pulsing outward, photons act much like a fluid source, and when pulsing inward, photons act much like a fluid sink.

Spatial Fluid (3.2, A1): A hypothesized compressible fluid that fills all space, and cannot be seen or felt although its effects can be observed and measured. Spatial fluid has properties similar to typical fluids such as air or water; it contains vortices, and supports density and pressure changes. Unlike the ether of the 1800's, the spatial fluid is here the source of all energy and matter.

Spatial Pressure (4.7, A8): The pressure in spatial fluid at any point.

Superconductivity and Electron Pairing (10.3, A4): Pairs of electrons, called Cooper pairs, are found in modern science to be intimately involved in superconductivity. Two types of electron pairing are discussed that could be candidates for Cooper pairs.

Time (12.6, A2): The idea of time seems to result from a finite speed of light that permits the sequencing of events. Here, time agrees with GPS results that show that time speeds up with altitude. Also, the speed of light increases here with altitude, suggesting that the rate of time may be a function of the speed of light, and conceivably could even be exactly proportional to the speed of light.

CPSIA information can be obtained at www.ICGtesting.com
Printed in the USA
LVOW061742190612

286825LV00001B/55/P